Elvia Patricia Valadez Sánchez

Thin film MOFs (SURMOFs) for application in gas separation

Thin film MOFs (SURMOFs) for application in gas separation

by

Elvia Patricia Valadez Sánchez

KIT Scientific Publishing

Dissertation, Karlsruher Institut für Technologie
KIT-Fakultät für Chemieingenieurwesen und Verfahrenstechnik

Tag der mündlichen Prüfung: 10. Oktober 2018
Erstgutachter: Prof. Dr.-Ing. Roland Dittmeyer
Zweitgutachter: Prof. Dr. Christof Wöll

Impressum

KIT Scientific Publishing

Karlsruher Institut für Technologie (KIT)
KIT Scientific Publishing
Straße am Forum 2
D-76131 Karlsruhe

KIT Scientific Publishing is a registered trademark
of Karlsruhe Institute of Technology.
Reprint using the book cover is not allowed.

www.ksp.kit.edu

Print on Demand 2019 – Gedruckt auf FSC-zertifiziertem Papier

ISBN 978-3-7315-0883-0
DOI 10.5445/KSP/1000088874

Thin film MOFs (SURMOFs) for application in gas separation

zur Erlangung des akademischen Grades eines

DOKTORS DER INGENIEURWISSENSCHAFTEN (DR.-ING.)

der KIT-Fakultät für Chemieingenieurwesen und Verfahrenstechnik des Karlsruher Instituts für Technologie (KIT)

genehmigte

DISSERTATION

von

M.Sc. Elvia Patricia Valadez Sánchez

aus Monterrey, N.L., México

Tag der mündlichen Prüfung: 10. Oktober 2018
Erstgutachter: Prof. Dr.-Ing. Roland Dittmeyer
Zweitgutachter: Prof. Dr. Christof Wöll

Dedicated to my parents, brother, extended family, friends, and my beloved.

To all of the Mexican women, I hope this work can be a source of inspiration of our amazing intellectual potential. Let's take advantage of the opportunities we have today thanks to the constant struggle and hard work of visionary women and men.

To the world, let us never settle down for as Einstein once said: "The important thing is not to stop questioning." Yeah, SCIENCE!

Dedicada a mis padres, hermano, familia extendida, amigos y a mi amor.

Para todas las mujeres mexicanas, espero este trabajo pueda ser una fuente de inspiración de nuestra impresionante capacidad intelectual. Aprovechemos las oportunidades disponibles hoy en día gracias a la continua lucha y trabajo arduo de mujeres y hombres visionarios.

Para el mundo, que jamás nos conformemos y en vez recordemos lo que Einstein en alguna ocasión dijo: "Lo importante es no dejar de hacerse preguntas". ¡VIVA LA CIENCIA!

Kurzfassung

Die Auftrennung von Alkanan und Alkenen ist derzeit nur durch energieaufwendige und kostenintensive Destillationsverfahren realisierbar. Um kommerziell attraktive Alternativen zu entwickeln, wurde in den letzten Jahrzehnten an verschiedenen Technologien geforscht. Membran-basierte Trennverfahren stellen dabei eine Gruppe der vorgeschlagenen Ansätze dar, die in den letzten 30 Jahren sehr intensiv untersucht und entwickelt wurden. Obwohl heute bereits viele Pilotanlagen für Membrananwendungen existieren, müssen nach wie vor große Herausforderungen überwunden werden, ehe eine breite kommerzielle Nutzung möglich wird.

Bei der Gastrennung werden verschiedene Arten von Membranen verwendet. Einer der kostengünstigsten und am meisten verwendeten Membrantypen sind dabei Polymermembranen. Die im Jahr 1991 von Robeson veröffentlichten Ergebnisse zur Korrelation zwischen der idealen Selektivität und der Permeabilität verschiedener Membranmaterialien stellen Richtwerte für die Beurteilung der Qualität von Membranen dar. Auf Membranmaterialien, die diesen Richtwert überschreiten wird daher ein besonderes Augenmerk gerichtet. Neben Molekularsieben oder Zeolithen werden in der jüngsten Zeit auch neue Materialklassen wie zum Beispiel metall-organische Gerüststrukturen (engl.: metal-organic frameworks, MOFs) als aussichtsreiche Membranmaterialien intensiv untersucht. MOFs sind mikroporöse und kristalline Materialien die aus metallischen Knotenpunkten und organischen Linkermolekülen aufgebaut sind. Im Vergleich zu anderen Materialien zeichnen sich MOFs durch eine Reihe herausragenden Eigenschaften aus. Je nach den verwendeten Metallen und/oder Linkermolekülen sind neben der

Größe und der Geometrie der Poren auch deren Reaktivität und Polarität exakt kontrollierbar. Die beliebige Kombination einer großen Zahl unterschiedlicher verfügbarer Linker und Metallknoten ermöglicht somit die Herstellung einer sehr großen Anzahl unterschiedlicher Gerüststrukturen. Diese Materialien besitzen eine hohe Permeabilität. In Bezug auf die MOF-Synthese sind bereits viele Methoden publiziert, so zum Beispiel die direkte solvothermale Synthese zur Herstellung von MOF-Pulvern oder das Wachsen von MOF-Filmen auf funktionalisierten Oberflächen. Um eine möglichst hohe Selektivität einer Membran zu erzielen, muss diese möglichst defektfrei hergestellt werden, was nach wie vor eine große Herausforderung darstellt. Die Klasse der ZIF-basierten (engl.: zeolitic imidazolate frameworks) MOFs – insbesondere ZIF-8 – zeichnet sich durch ihre hohe thermische und mechanische Stabilität aus. Aufgrund der sehr kleinen Porenfenster (ca. 0,34 nm) und Hohlräume (ca. 1,1 nm), die im Bereich des kinetischen Durchmessers verschiedener Gasmoleküle liegen, stellt ZIF-8 den am meisten untersuchten MOF für die Gastrennung dar. Struktur- und Sorptionsuntersuchungen haben gezeigt, dass der Orientierung der ZIF-Kristalle eine hohe Bedeutung für den erleichterten Zugang der Gastmoleküle in die Poren und damit für eine gute Membranleistung zukommt. Eine Möglichkeit hoch-orientierte, defektfreie und reproduzierbare Membranen herzustellen ist dabei ein schichtweises epitaktisches Wachstum aus der Flüssigphase. Bei diesem Ansatz lässt man auf einem chemisch vorkonditionierten Substrat die MOFs orientiert aufwachsen, wobei die Schichtdicke der MOFs durch die Zahl der Abscheidezyklen bestimmt wird und damit sehr genau eingestellt werden kann. Diese oberflächen-verankerten MOFs werden als SURMOFs (engl.: surface-anchored MOFs) bezeichnet. Auf diese Weise konnten z.B. die SURMOFs HKUST-1 oder ZIF-8 hergestellt werden.

Die Bewertung der Leistungsfähigkeit der Membranen in Hinblick auf ihre gastrennenden Eigenschaften gegenüber Alkanen und Alkenen war Schwerpunkt der vorliegenden Doktorarbeit. Dazu wurden zunächst die experimentellen Parameter für ein automatisiertes layer-by-layer (LBL) Tauchbeschichtungsverfahren zur Herstellung eines ZIF-8 SURMOFs auf porösen α-Aluminiumoxid-Trägersubstraten (mit und ohne Goldbeschichtung) getestet und optimiert. Dazu wurden Zinknitrathexahydrat als Metallquelle und 2-Methyl-1H-Imidazol als Linkermolekül verwendet. Die Zahl der Abscheidezyklen variierte zwischen 100, 125, 150, 175 und 200. Um die ZIF-8 Membranen hinsichtlich ihrer Qualität zu charakterisieren, wurden verschiedene Analysenmethoden angewendet: so wurde mit Hilfe der Röntgendiffraktometrie (engl.: X-ray diffraction, XRD) die Kristallinität sowie die Wachstumsrichtung der MOF-Membranen untersucht, während die Infrarot-Reflektions-Absorptionsspektroskopie (IRRAS) zur Bestimmung der chemischen Zusammensetzung sowie die Rasterelektronenmikroskopie zur Bestimmung der morphologischen Eigenschaften der Oberfläche zum Einsatz kamen.

Anschließend wurde die Qualität der Membranen hinsichtlich etwaiger Oberflächendefekte oder Risse mit Hilfe druckabhängiger Permeationsexperimente untersucht. Für diese Versuche wurden die Membranen in mittels 3D-Druck gefertigten Modulen montiert wodurch ein gleichmäßiger Gasfluss über die gesamte Membranoberfläche sichergestellt werden konnte. Zusätzlich erlaubten diese Permeationsexperimente die Bestimmung der erforderlichen Mindestanzahl an Abscheidezyklen, um leistungsfähige Membranen herzustellen. Nachdem die experimentellen Bedingungen für die LPE LBL-Herstellung der ZIF-8 SURMOF-Membranen festgelegt war, wurde deren mechanische sowie thermische Stabilität untersucht, da es zwar viele Publikationen zu Stabilitätsuntersuchungen für Pulver-Materialien gibt, nicht jedoch für dünne ZIF-8 Membranen.

Nachdem die Stabilität der ZIF-8 Membranen unter Umgebungsbedingungen sichergestellt werden konnte, wurde mit der Durchführung von Permeationsexperimenten zur Gastrennung begonnen. Diese lieferten zuverlässige experimentelle Daten, die anschließend für Simulationszwecke verwendet wurden. Entsprechend dem erarbeiteten Syntheseprotokoll wurden diverse Membranen hergestellt, um anschließend ihr Anwendungspotential hinsichtlich der Trennungseigenschaften für binäre Gasgemische mittels der Wicke-Kallenbach-Methode zu testen und zu beurteilen. Zusätzlich wurden Permeationsexperimente mit einzelnen Gasen durchgeführt, um die Oberflächenbedeckung in der Retentat- und der Permeat-Kammer des 3D-Moduls zu ermitteln. Diese experimentellen Daten wurden anschließend mit Daten zur Adsorption für ZIF-8 Systeme aus der Literatur kombiniert, um die Maxwell-Stefan (M-S) Oberflächendiffusivität der Gase zu berechnen, die für die Simulationsstudie erforderlich waren. Die resultierenden experimentellen Daten der Gastrennungsversuche wurden dabei als Grundlage für MATLAB-Simulationen des Gastransports durch die Membran und den Träger basierend auf dem von Krishna erweiterten Maxwell-Stefan Modell für Binärgasmischungen verwendet. Das Modell beschreibt die Diffusion als Oberflächentransportphänomen von binären Gasmischungen auf der Nanometerskala und beinhaltet die molekularen Adsorbat-Adsorbat Wechselwirkungen sowie die Wechselwirkung zwischen den Gastmolekülen und den Porenwänden und berücksichtigt somit auch Adsorptionskräfte. Für die Simulationen musste zudem ein geeignetes Adsorptionsmodell gefunden und anschließend getestet werden. Zwei Modelle wurden verwendet: 1) IAST (engl.: ideal adsorbed solution theory) sowie 2) das erweiterte Langmuir-Modell. Die Simulationen basieren auf einer Diskretisierung der ZIF-8 SURMOF Schicht, die eine Bestimmung der molekularen Anteile jeder Komponente durch die Membran ermöglicht. Der Einfluss der α-Aluminiumoxid-Träger auf die Transportvorgänge bei der Gastrennung wurde auch untersucht. Mit den aus der Simulation erhaltenen Daten wurde ein besseres Verständnis der Transportvorgäng innerhalb der mit dem schichtweisen epitaktischen Wachstum aus der Flüssigphase hergestellten ZIF-8 Membranen erreicht.

Abstract

Alkane/alkene separation is currently characterized by energy-intensive low temperature distillation processes, requiring large equipment and areas, resulting in high capital and operation costs. Pursuing the development of more commercially attractive alternatives, various technologies have been studied. Membrane-based separation processes are one of the proposed approaches that has been intensively investigated and developed over the last 30 years. Though several pilot plants exist nowadays, this technology still exhibits great challenges that need to be overcome to achieve commercial feasibility.

When speaking about gas separation, diverse types of membranes have already been developed. One of the largest and cheapest studied groups corresponds to polymeric membranes. In 1991, Robeson determined the relationship between permeability and selectivity in polymeric membranes establishing their trade-off bound; in order to exhibit promising results towards the possibility of industrial application, such bound must be overcome. Seeking to exceed polymeric membranes' performances and therefore the stated bound, the development of new technologies began including carbon molecular sieves, zeolites, and more recently MOFs. All of these showed several promising features that further deepened their investigation.

MOFs are porous solid materials composed of metal nodes and organic linkers bonding together forming crystalline structures. When comparing MOFs to other established technologies, some exclusive features can be identified. For instance, depending on the linker and the metal chosen, the control over the pore geometry, i.e., size and shape, and pore properties such as reactivity and polarity is possible. Also given their building blocks, the possible number of combinations is staggering. Furthermore, these materials have shown the ability to deliver simultaneously high permeability and selectivity.

When speaking about MOF synthesis, a great amount of methods have been published including direct synthesis, secondary growth, surface functionalization methods, among others. Unfortunately, the synthesis of defect-free membranes is highly challenging, but of great importance for the achievement of highly selective membranes. One of the most famous subclasses within MOFs are zeolitic imidazolate frameworks, ZIFs, which are known for their thermal and mechanical stability. Among them, ZIF-8 has been one of the most studied frameworks within gas separation applications due to its large inner cavities enclosed by narrow sized windows, ranging within the kinetic diameters of various gas molecules.

According to structural and sorption studies, achieving highly oriented ZIF-8 crystals is of great importance in order to facilitate the access of guest molecules, enhancing the membrane's performance. One of the approaches that seeks to achieve highly oriented, defect-free, and reproducible membranes is the liquid phase epitaxial (LPE) layer-by-layer (LBL) technique. This approach incorporates the use of interfaces in order to initiate and control the growth of MOFs in terms of orientation and crystallinity. The thickness of the SURMOF can be exactly adjusted with the number of deposition cycles; various studies have been conducted delivering very promising results, for instance HKUST-1 and ZIF-8 systems have been done previously.

The evaluation of the aforementioned approach and its performance in alkane/alkene gas separation was the core point of study of this PhD work. To begin, the experimental parameters for the automated dipping LPE LBL ZIF-8 SURMOF deposition were tested and optimized on two different porous α-alumina supports (non-coated and Au-coated); metal nodes: zinc dinitrate hexahydrate, organic linkers: 2-methyl-1H-imidazole. Different deposition cycles (100, 125, 150, 175, and 200) were tested for the membrane synthesis, and characterization studies by x-ray diffraction (XRD), infrared reflection absorption spectroscopy (IRRAS), and scanning electron microscopy (SEM) were done in order to evaluate the membranes' quality regarding crystallinity and growth orientation, chemical composition, and surface morphology, respectively. Furthermore, the membranes' overall quality was evaluated via H_2 pressure-dependent experiments. For these tests, the membranes were firstly placed inside specifically designed 3D-printed modules that ensured an even gas flow over the surface. These experiments also determined the minimum number of deposition cycles required in order to ensure a fully covered support that could deliver higher selectivities.

Once the LPE LBL ZIF-8 SURMOF system was established, the material was further analyzed regarding its mechanical and thermal stability. Though ZIF-8 has been thoroughly characterized in its powder form, when discussing thin film stability, conditions can be significantly affected and currently limited data exists. Furthermore, by confirming the stability of the system at determined operating conditions, permeation and gas separation tests could be successfully performed delivering reliable experimental data that was later on used in simulation runs.

Several membranes were then synthesized according to the determined number of deposition cycles required and further tested in gas mixture separation using the Wicke-Kallenbach method, which permitted to assess their application potential in that field. Additionally, single feed gas permeation experiments were conducted in order to obtain surface coverage data on the retentate and permeate chambers. These experimental data was then combined with adsorption literature for ZIF-8 systems in order to calculate the Maxwell-Stefan (M-S) surface diffusivities, that were required in the simulation study.

Finally, the resulting experimental data was then used to perform simulation runs for the performed gas separation experiments using a MATLAB® program based on the generalized M-S surface diffusion model proposed by Krishna (1990), which is one of the most used for transport of gases in microporous solids. This approach analyzes the diffusion of gas mixtures at the nanopore level assuming surface transport. The model comprises of the molecular interactions in the

SURMOF membrane, taking into account the adsorbate-adsorbate interactions, as well as the interactions between the molecules and the adsorption sites. The simulation also requires the selection of an adsorption model. Two different adsorption models were tested: IAST and extended Langmuir. The approach followed consisted of a discretization of the ZIF-8 SURMOF layer in order to determine the molar fractions of each of the components along the framework; the support's influence was also considered. With the obtained simulated data, a better understanding of the diffusion mechanism through the ZIF-8 SURMOF framework, as well as the role different parameters play during it was achieved.

Acknowledgments

First of all, I would like to deeply thank both of my supervisors, Dr. Katja Haas-Santo and Dr. Hartmut Gliemann for all of the support in these last 3 years. Your guidance was a pillar in this accomplishment and your advice was always highly valuable. I would also like to thank Prof. Dr.-Ing Roland Dittmeyer and Prof. Dr. Christof Wöll for all of the advice and encouragement provided. Our meetings were always very motivating, suggesting constantly ways to enrich my work and achieve better results, helping me become a better scientist. Thank you for the mentoring and above all for trusting and giving me the opportunity of further developing myself at the Karlsruhe Institute of Technology. I am and will always be sincerely grateful for these years at "Campus Nord", where the early stage of my professional life began. A special thanks also to Prof. Dr.-Ing. Achim Dittler and Prof. Dr.-Ing. Hermann Nirschl for their time and consideration of this PhD work.

I would also like to thank all of my colleagues at IFG, IMVT, and other institutes. I had a really nice time in these last 3 years, not only at work, but also in diverse after-office events we had the opportunity to organize. A special thanks to Dr. Sonia Escolástico, Edgar Hansjosten, Jonas Wohlgemuth, Dr. Wenjin Ding, Kay Dyrda, Ulrich Schygulla, Dr. Wencai Zhou, Dr. Nicolás Giraudo, Dr. Sreeja Nair, Fabrice Layé, Ángel González, and Daniela Sánchez. Your help was key in this accomplishment and I will always be thankful for all those hours we shared, for your advice, and thoughtful insights.

Furthermore, I sincerely appreciate the great financial support provided by the Helmholtz Graduate School Energy-Related Catalysis. The soft skills courses, conferences, summer schools, winter schools, and social events helped me develop a greater set of skills that will clearly shape my future. A special thanks to Dr. rer. nat. Claudia Antinori and Dr. rer. nat. Steffen Tischer for all the coordination within this group.

Last but not least, I would like to thank my parents for all the tools they have provided me with that have helped me along my personal and professional development. I will never be able to compensate everything you have given to me, but I am deeply grateful and hope to make you proud every day of my life. I would also like to thank my boyfriend, brother, extended family, and friends. Doing a PhD in Germany has been one of the greatest and most challenging experiences in my life and it wouldn't have been the same without you. Either close or far, your love and care permitted me to carry on in good and bad times. Thanks for your unconditional support and for pushing me to become not only a better scientist, but also a better person. Your love guides me every single day.

Finalmente, quisiera darle las gracias a mis papás por haberme brindado las herramientas que me han ayudado a lo largo de mi desarrollo personal y profesional. Jamás podré recompensar todo lo que me han dado, pero estoy profundamente agradecida y espero llenarlos de orgullo cada día de mi vida. Igualmente quisiera darle las gracias a mi novio y a su familia, a mi hermano, a mis abuelos, a mis tías, tíos, primos y amigos. Mi doctorado en Alemania ha sido una de las experiencias más increíbles y retadoras en mi vida y definitivamente no hubiera sido lo mismo sin ustedes. Ya fuera cerca o lejos, su amor y cuidado me permitieron sobrellevar momentos prósperos y adversos. Gracias por su apoyo incondicional y por impulsarme a convertirme no sólo en una mejor científica, sino también en una mejor persona. Su amor me guía cada día.

Karlsruhe, August 2, 2018 Elvia Valadez

Contents

1. Introduction

1.1. Motivation

Alkane/Alkene separation a.k.a. paraffin/olefin separation represents an important industrial operation primarily performed via an energy-intensive low-temperature distillation process [1]. Due to their similar volatilities and molecular sizes, the separation process is energy intensive requiring columns with heights of up to 90 m with 150-200 trays and involving temperatures and pressures between 183-233 K and 16-20 bar, respectively [2, 3], being therefore credited as the most intensive commercially practiced distillation [4]. Based on such high capital and operation costs, diverse alternative approaches have been proposed and studied including membrane separation. In the last 30 years, over 2000 articles and patents have been and continue to be published regarding hydrocarbon separation and removal, indicating an intense effort in this field, still presenting challenges to be overcome before becoming commercially attractive [4, 5].

Pursuing gas separation, different groups of membranes have been studied including polymeric membranes. In 1991 [6], Robeson published a selectivity versus permeability plot with the data from these membranes lying on or below a straight line defined as the upper bound trade-off curve, later known as Robeson's bound, which theoretically showed the bound that must be overcome in order for the membranes to be considered good performing from a commercial point of view. Based on this, carbon molecular sieves (CMS) started to be investigated exhibiting good potential, but being difficult to process, expensive, and inherently fragile [7].

Further new and reemerging technologies surpassing this bound are still being studied, including zeolites and metal-organic frameworks (MOFs) which have been proven to deliver simultaneously high permeability and selectivity [4, 8]. When comparing MOF membranes to other technologies, several promising features can be identified including a wider range of pore sizes, pore shapes, and a higher degree of control over pore functionality at the molecular level [9, 10].

The synthesis of MOFs include a wide range of published procedures including direct synthesis, secondary growth, and surface functionalization methods, among others [11]. One of the approaches consists in the use of interfaces to initiate and control the growth of MOFs in terms of orientation and crystallinity via liquid phase epitaxial (LPE) layer-by-layer (LBL) deposition, e.g., dipping, spraying, and spin-coating. MOFs grown that way are called surface-anchored metal-organic frameworks (SURMOFs). This technique has been proven to deliver highly oriented crystals [12, 13, 14, 15] overcoming also the so-called surface barriers [16], frequently encountered by MOF powder particles resulting in well-suited candidates to be further studied in gas separation. Zeolitic imidazolate frameworks (ZIFs) are the MOF subclass most thoroughly studied for this application [17], one of the reasons being their known high thermal stability [18, 19]. For these reasons, growing this material via the aforementioned approach and evaluating its performance in alkane/alkene gas separation was the core point of study of this research work.

1.2. State of the art

1.2.1. MOFs

Metal-organic frameworks (MOFs) are a class of porous solid materials composed of two types of building blocks: organic linkers and metal nodes as depicted in Fig. 1.1. The interaction between these last two results in the formation of crystalline networks. Usually, a one-pot synthesis is carried out in order to prepare MOFs as bulk materials, i.e., both building blocks are mixed in a solvent and treated under solvothermal conditions leading to the precipitation of MOF crystals [20, 21]. The length and the chemical functionalization of the chosen linker molecules permits a high control and flexible tailoring of the geometry (shape and size) and the properties (polarity and reactivity) of the pores, respectively [22]. Given the vast number of feasible combinations of the available building blocks, over 70,000 different MOF materials in bulk form have been discovered so far and have been investigated for application in diverse fields including hydrogen storage, gas separation, carbon dioxide capture, catalysis, smart sensors, and drug delivery [23].

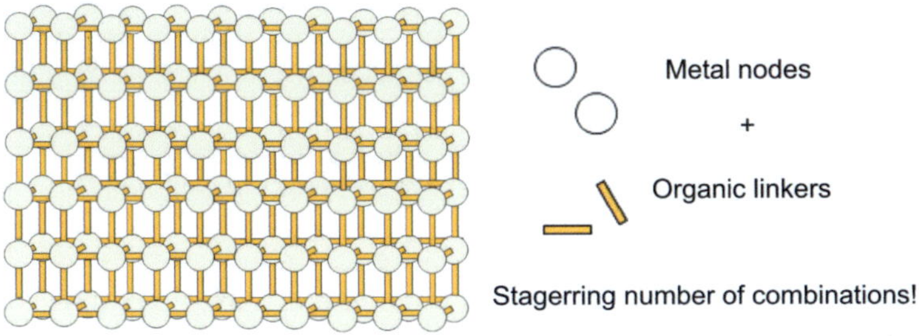

Figure 1.1: MOF schematic and its building blocks

Prior to the investigation of these materials, a lot of resources were destined towards investigation into zeolites. An important point of this fact is that the tools designed for this area of research, e.g., seeding, microwave heating, ceramic porous supports, etc., can be also applied for the synthesis of MOF membranes [18] facilitating, economically speaking, the deep exploration that this field is being experiencing.

To differentiate MOFs from other porous materials, interesting promising features have been studied and confirmed along the past 20 years. Some of their excelling properties include [24, 25, 26, 27]:

- World records in surface area of up to 6500 $m^2 \cdot g^{-1}$
- Combination of flexible and robust frameworks: some MOFs possess permanent porosity and robust frameworks comparable to zeolites, while others are classified as flexible or capable of undergoing stimuli-driven structural rearrangements for instance via temperature and pressure changes or due to the presence of guest molecules
- High mobility of guest species in the frameworks
- Staggering number of novel MOF chemical compositions

Still challenges exist for MOFs, regarding primarily mechanical and thermal stability, perceived as highly problematic. Nonetheless in recent years, the list of thermally and water-stable MOFs has increased, expanding significantly their application areas including the possibility of being considered as effective heterogeneous catalysts [28].

1.2.2. SURMOFs

Even though the use of MOFs in their bulk form is possible for the previously mentioned applications, when speaking about the synthesis of gas permeable membranes, the achievement of well-intergrown continuous defect-free membranes is not an easy task. In order to have high quality membranes, pinholes, grain boundary defects, or intracrystalline and intercrystalline cracks must be completely eliminated seeking the minimization of non-selective gas transport which would otherwise considerably affect the membrane's performance [29, 30]. Furthermore, MOF powder particles tend to suffer from so-called surface-barriers [16], which critically hinder the diffusion of gases. Great efforts have been therefore done in order to accomplish such high performing membranes, and a lot of different techniques have been tested in order to determine an optimal path, e.g., in-situ seeding and secondary growth [31]. Despite the amazing pioneering work and achievement in diverse cases of high-quality membranes, several issues are yet to be overcome, for instance reproducibility and more importantly practical application of membranes in separation processes [29].

For this reason, new approaches that permit a more controlled MOF growth have been sought, seeking to deliver higher quality coatings or membranes. One of them includes the so called surface-anchored metal-organic frameworks (SURMOFs). SURMOFs are grown by utilizing interfaces to initiate and control the growth of MOFs in terms of orientation and crystallinity. First of all a self-assembled monolayer (SAM) is grown on the substrate. Afterwards, the SURMOF is synthesized by a liquid phase epitaxial (LPE) layer-by-layer (LBL) deposition procedure [12], e.g., by dipping [13], spraying [32], or spin coating [33]. As a result, monolithic SURMOFs exhibiting a high degree of crystallinity are obtained. The well-suited properties of these materials make them highly attractive for membrane applications. To depict the SAM and SURMOF formation, Fig. 1.2 is presented next.

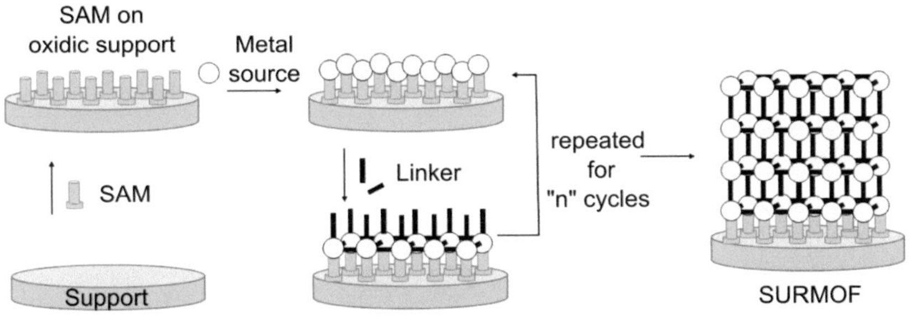

Figure 1.2: SAM and SURMOF formation

1.2.2.1. SAM formation

Before initiating the LBL SURMOF deposition, seeking the optimization of the adhesion of the SURMOF, the supporting substrate is chemically activated. The way of surface activation depends on the metal source and linker molecules used for the SURMOF synthesis, as well as on the substrate material, affecting also the orientation of the SURMOF layer on the substrate. For oxidic substrates, activation by oxygen plasma treatment, UV-light irradiation (formation of -OH groups) or by reaction with appropriately terminated silanes to form a SAM is possible. Regarding gold-coated substrates seeking to enhance a controlled growth, the functionalization is achieved via thiol-based SAMs. This activation provides an excellent organic template for further highly oriented SURMOF growth [24].

1.2.2.2. Liquid phase epitaxial layer-by-layer deposition procedure by dipping

LPE consists in the sequential treatment of the substrate with the solutions of the MOF building units and the removal of the non-coordinated components after each step [14] performed at normal temperature and pressure. In Fig. 1.3 the LPE LBL deposition for SURMOFs via dipping is depicted. As it can be observed, a sample holder is submerged into different beakers with different solutions in a determined sequence. The concentrations of the solutions, as well as the deposition temperature and times depend on the MOF to be grown and have to be optimized for each case through experimentation. Important also to the technique is the removal of unreacted components after each step, achieved through the rinsing of the sample in pure solvent. This process of submersion is repeated for several times, or cycles, depending on the desired thickness. Through this method, the fabrication of continuous and highly oriented MOF films with controllable thickness on an –OH terminated SAM on a non-porous gold surface has already been proven [23]. Concisely, the building blocks selected determine the properties and size of the pores [22], the growth orientation of the SURMOFs is controlled by the chemical termination of the substrate, and the film thickness may be adjusted exactly, due to the proportionality of layer thickness to the number of growth cycles [24, 34].

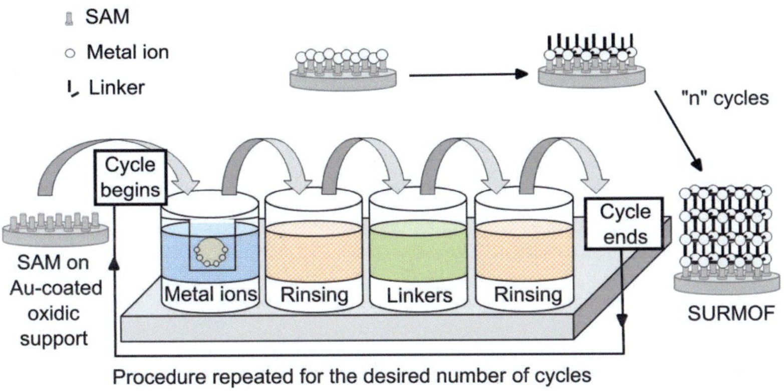

Figure 1.3: Scheme of the LBL deposition of SURMOFs (adapted from [35])

1.2.3. ZIFs

Zeolitic imidazolate frameworks (ZIFs) are a subclass of MOFs consisting of metal centers tetrahedrally coordinated by nitrogen atoms of the imidazolate bridging ligand. Their analogous formation to the Si-O-Si angle in silicas and zeolites, makes them share zeolitic traits permitting a combination of chemical and thermal stability, together with the exclusive features MOFs possess regarding topological diversity and porosity tunability; 90 types of structures have been discovered so far [36].

This subclass is also recognized as the most thoroughly studied in the gas separation field [17], based on several important features they possess. Their high thermal stability is one of the most recognizable, being able to withstand operation conditions of up to 550 °C in N_2 [19] and up to 200 °C in 3 vol.% steam [18]. Additionally, experimental results and simulations of ZIF-8 have suggested the framework's flexibility [17, 37]. Moreover, Huang et al. [10] recently reported ZIF membranes to exceed the known Robeson "upper-bound" for polymeric membranes. Such traits together with their nanometer size pores have made ZIFs attractive candidates in the field of membrane-based gas separation [17].

1.2.3.1. ZIF-8 characterization

Within the ZIF subclass, ZIF-8 is one of the most studied. Based on crystallographic data, it has been determined to be composed of pore size windows of 3.4 Å in diameter enclosing a large cavity of 11.1 Å in diameter [13], making ZIF-8 a highly porous and open framework possessing large accessible pore volume. Likewise, such nanometer-size pores ranging within the kinetic diameter of several gas molecules has added particular interest for ZIF-8 within gas separation applications [38]. Besides this field of study, ZIF-8 catalytic applications [39] and sensor development [40] have been also reported in the recent past. The crystal structure of ZIF-8 is shown in Fig. 1.4.

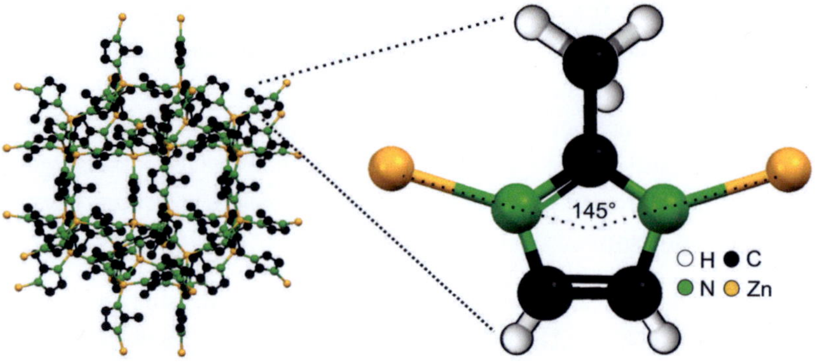

Figure 1.4: Crystal structure of ZIF-8 and its building units [13, 41]

ZIF-8 is characterized by a sodalite (SOD) topology. As shown in Fig. 1.5, its structure is composed of 4- and 6-membered rings which can be oriented for instance, in the (100) or (110) planes. According to sorption and structural studies, the entry of guest molecules is an exclusive task of the six-membered rings [23], a reason for which a parallel alignment in the (110) plane is highly desired. Achieving such orientation should permit a facilitated entrance and diffusion of guest molecules through the pores, instead of forcing them to travel through a zigzag path [31].

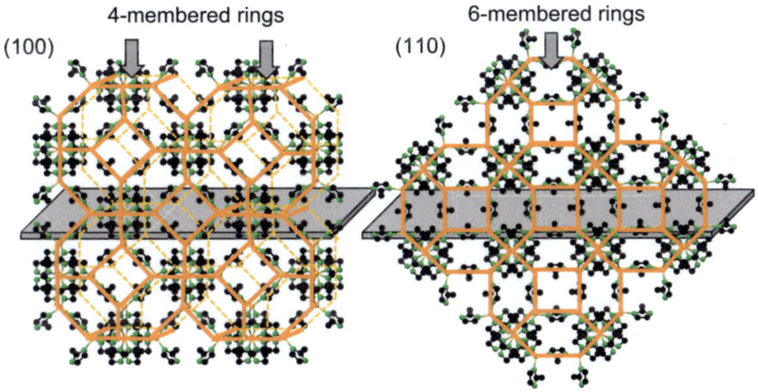

Figure 1.5: ZIF-8 four- and six-membered rings oriented on the (100) and (110) planes (based on [23])

X-ray diffraction (XRD) Due to the gas separation application sought in this study, a highly oriented crystal film is of major importance. In order to investigate this and reach the smallest building blocks of the film, i.e., unit cells, the use of x-rays is necessary.

Diffraction patterns are obtained when an incident beam passes through a crystalline material. With the help of the diffracted beam, as well of the transmitted beam, the diffraction angle (2θ) can be obtained. Through this procedure and based on Bragg's Law [42]

$$n\lambda = 2d \sin\theta \tag{1.1}$$

a diffraction pattern can be recorded. Diffractograms are composed of two axes, x being the diffraction angle 2θ and y being the intensity. The peaks' positions are determined by the first one, while the intensity is determined by the latter. Thanks to modern user interfaces, indexing, i.e., the process of determining the unit cell dimensions from the peak positions assigned by Miller indices referring to the plane on which the unit cells are oriented, also known as hkl, has been facilitated [43]. With the obtained information the users are then able to analyze the data and determine if the measured sample has a defined orientation or not, i.e., if it is really growing as an oriented crystal or if it is a non-oriented powder-like coating.

As mentioned previously, ZIF-8 is characterized by its sodalite topology [13]. This SURMOF belongs to the space group $I\bar{4}3m$, standing for a body-centered lattice type. Its reported cell lengths range between 16.9 Å and 17 Å with 90° angles (α,β,γ), and a cell volume of around 4,900 Å^3 [13, 19, 44]. Other general characteristics from ZIF-8 have been thoroughly studied in the past and are summarized in Tab. 1.1; the data was extracted from the reported work by Krishna and van Baten [44].

Structure	a/	b/	c/	Unit cell volume	Pore volume per unit cell	Fractional pore volume	Pore volume	Framework density
	Å	Å	Å	Å^3	Å^3	ϕ	[cm^3·g^{-1}]	[kg·m^{-3}]
ZIF-8	16.99	16.99	16.99	4905.2	2336.97	0.476	0.515	924.25

Table 1.1: Unit cell dimensions and other characteristics of ZIF-8. Table created by author from data in [44].

As reported in literature [23, 45], the XRD pattern of ZIF-8 is characterized by peaks in the (110), (200), (211), (220), (310), and (222) planes within a range of 2θ= 5.5–20°. Nonetheless, based on the sorption and structural studies already mentioned, the achievement of a predominant orientation on the (110) plane is of high interest for this work. A powder diffractrogram pattern of ZIF-8 extracted from Lewis's published work [46] can be found in Appendix A.1.

Infrared reflection absorption spectroscopy (IRRAS) Surface chemical composition of the ZIF-8 film is another important point that must be studied as part of the material's characterization. One of the possible techniques is infrared (IR) spectroscopy, which analyzes the chemical composition of materials through the measurement of the IR light that is absorbed by a substance due to the excitation of molecular vibrations, the so-called vibrational modes. As different modes are excited by light of different wavelengths, the absorption bands of an IR spectrum give hints of which molecular groups are present in the substance under investigation. A mode can be identified by IR spectroscopy, i.e., it is IR active, if the following selection rule is satisfied: vibrationally originated dipole moment changes in the corresponding molecular group must take place. Otherwise this mode is IR-inactive (e.g., H_2 and O_2) [47, 48].

In an IR spectrum, the absorption is plotted as a function of the wavenumber (i.e., the reciprocal of the corresponding wavelength). This is due to the fact that frequency is proportional to the energy and since the wavelength is inversely proportional to the frequency, it is also inversely related to the energy [49]. For this reason, another unit proportional to the energy termed wavenumber, $\bar{\nu}$, is commonly used in IR spectroscopy.

To characterize thin films or coatings on metal surfaces which are not transparent, infrared reflection absorption spectroscopy (IRRAS) can be used; the surface chemical composition of the ZIF-8 SURMOF membranes in this work will be characterized through it. This technique is based on three important points. First of all, in order for IR light to be absorbed, the molecule must have a (portion of the) dipole moment change perpendicular to the surface, otherwise vibrations cannot be excited and it is consequently termed IR inactive [48]. Second, only the parallel polarized light (p-polarized light) with respect to the plane of incidence is absorbed by thin layers on metal substrates which is due to the phase shift according to the studies of R. Greenler in 1965 [50]. Finally, and also based on this last point, the strongest absorption is achieved at a defined light incidence angle, which is a function of the metal. For gold, the maximum is achieved at an incidence angle of 88°.

According to reported literature, ZIF-8's composition can be described based on the origin of its bonds, given the fact that the majority of its absorption bands are associated with the vibrations of the imidazole units [51]. The characteristic ZIF-8 absorption bands are reported in Tab. 1.2 [45, 51, 52].

IR Band [cm^{-1}]	Assignment
1100-1400	C-N absorption bands
1384	Methyl bending
1584	C=N stretch
2929	Aliphatic C-H stretch (imidazole)
3135	Aromatic C-H stretch (imidazole)

Table 1.2: Characteristic IR absorption bands in ZIF-8 [45, 51, 52]

1.2.4. SURMOF membranes for gas separation

Separation techniques have existed since hundreds of years. When speaking about paraffins a.k.a. alkanes and olefins a.k.a. alkenes separation, cryogenic distillation is one of the most common procedures followed. Nonetheless, being this an energy intensive and therefore expensive procedure, alternate techniques have been sought. When speaking about membrane-based separation as a possible choice and comparing it to other established technologies, several positive aspects can be identified: lower energy requirements and investments [17, 18]. Nonetheless, the development of highly efficient membranes, regardless the wide variety that exist, poses still a key issue primarily regarding its optimization challenge in terms of selectivity for specific gases.

Membranes with a more or less wide pore size distribution are applied in the separation of molecules of different sizes. Unfortunately, the majority of the membrane materials are usually not flexible or capable of tuning an exact pore size, pore shape, or chemical properties of the pores, e.g., polarity and reactivity. Metal-organic frameworks (MOFs) are a material class that offers the possibility of covering a wider range of pore sizes, pore shapes, and a higher degree of control over pore functionality at the molecular level [9, 10].

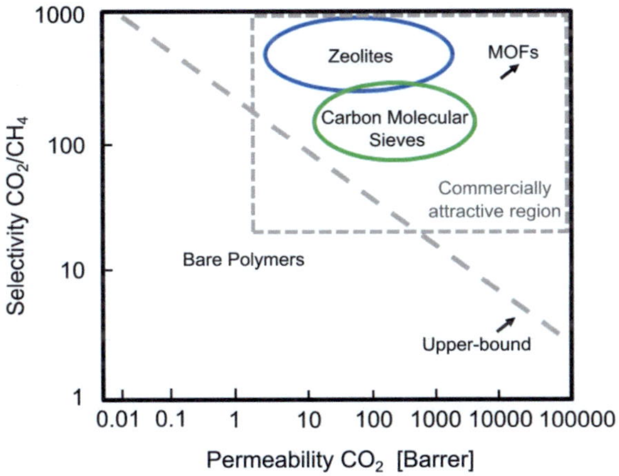

Figure 1.6: Robeson's plot overview for various membrane types for the CO_2/CH_4 separation. Figure recreated by author from data in [8].

Worthy to mention is the trade-off that still exists in membranes today. According to scientist Lloyd Robeson and his paper published in 1991 [6], a direct trade-off between permeability and selectivity exists, meaning highly selective membranes have usually low permeabilities, i.e., mostly no separation is being performed since compounds are not going through, and lowly selective membranes have high permeabilities, i.e., a poor separation is being performed since most components are flowing through it. This "bound" he defined was updated in 2008 [53], but still great challenges exist for membranes today. Fig. 1.6 adapted from Gascon's work [8] shows a summary for diverse membrane types comparing these two parameters. As it can be observed, MOFs have been reported to surpass the Robeson's trade-off bound becoming interesting and potential candidates for membrane materials [9, 10]. When speaking about prices, polymeric membranes happen to be quite cheap ranging around 20 €/m^2 [54], while inorganic membranes being more selective are quite expensive ranging around 2000 €/m^2 [54, 55].

These different types of membrane materials are also related to a very important parameter: pore size. Depending on the gases to be separated, the pore size of the membrane must be considered, since this will play a crucial role in defining the transport mechanism to be presented along the membrane.

1.2.5. Mass transport mechanisms in porous membranes

According to the International Union of Applied Chemistry (IUPAC) three different pore size categories exist [56]:

- Macropores (larger than 50 nm)
- Mesopores (between 2 nm and 50 nm)
- Micropores (less than 2 nm)

The classification was performed based on the different types of forces controlling the adsorption performance within pores of these sizes [56].

Furthermore, various transport mechanisms in the gas phase have been identified and defined in porous media. Relevant to this study are viscous flow, molecular diffusion, Knudsen diffusion, and surface diffusion.

An important term which is discussed within the regimes' definitions is the mean free path, λ_m, defined as the average distance traveled by a molecule between successive collisions. The mean free path can be calculated using the following equation [57]:

$$\lambda_m = \frac{1}{\sqrt{2} n \pi \sigma^2} \tag{1.2}$$

where n is the number of molecules per unit volume and σ is the diameter of the ideal gas spherical molecules. A summary of the aforementioned mechanisms is shown in Tab. 1.3, and further details of every mechanism are presented next.

1.2.5.1. Viscous flow

Viscous flow is the mass transport mechanism occurring in cases where the pore diameter of the material is larger than the mean free path and a significant total pressure difference exists between the two sides of the membrane, i.e., the driving force for these mechanism is an absolute pressure gradient [58, 59, 60]. The flux in this pressure-driven regime can be estimated depending on model assumptions, e.g., pore shape. Given cylindrical pores, the Hagen-Poiseuille equation can be used [61, 62]:

$$J_i^v = \frac{\varepsilon_m r_p^2 \Delta P}{8 \eta \tau_m \delta_m} \tag{1.3}$$

where J_i^v is the viscous molar flux, ε_m is the membrane's porosity, r_p is the average pore radius, ΔP is the pressure difference across the membrane, η is the viscosity of the gas, τ_m is the tortuosity of the membrane, and δ_m is the membrane's thickness.

1.2.5.2. Molecular diffusion

In the absence of a total pressure difference and pore diameters larger than the mean free path, mass transport is governed by molecular diffusion. In this mechanism, the number of molecule-molecule collisions is much higher than that between the molecules and the walls [63] which are driven by concentration gradients [58, 59, 60]. Given a binary mixture of i and j,

$$J_i^m = -D_{ij}^m \nabla C_i \tag{1.4}$$

where J_i^m is the molar flux by molecular diffusion, D_{ij}^m is the binary diffusivity, and ∇C_i is the concentration gradient [64, 65]. The binary diffusivity can be calculated using the empirical correlation developed by Fuller, Schettler, and Giddings [66].

1.2.5.3. Knudsen diffusion

As the diameter of the pores, d_p, starts decreasing, this parameter might eventually become smaller than λ_m. When that occurs, molecules begin to collide more frequently with the walls than with other molecules, entering then the Knudsen diffusion regime. A parameter to help determine whether such regime has been reached is the Knudsen number, K_n, defined as the division of the mean free path by the diameter of the pore, d_p [56, 67]:

$$K_n = \frac{\lambda_m}{d_p}. \tag{1.5}$$

As a common reference, a K_n above 1 is considered to be in the Knudsen diffusion regime, while below 1 is considered to be in the viscous flow regime. Nonetheless, for Knudsen numbers between 1 and 10, a mixture of Knudsen diffusion and viscous flow is usually present [5]. As a rule of thumb, systems with a $K_n > 10$ present Knudsen diffusion as the dominant transport mechanism [68]. The following model permits the estimation of the molar flux based on a concentration gradient as the driving force [69],

$$J_i^K = -D_i^K \nabla C_i \tag{1.6}$$

where J_i^K is the Knudsen molar flux, D_i^K is the Knudsen diffusivity, and ∇C_i is the concentration gradient of species i. Furthermore, the D_i^K is based on the momentum conservation arguments done by Knudsen and later refined by Smoluchowski with respect to molecular trajectories [70]. The D_i^K in the gas phase for a cylindrical pore can be calculated through the following equation [64, 71]:

$$D_i^K = \frac{d_p}{3}\sqrt{\frac{8RT}{\pi M_i}} \tag{1.7}$$

which can be further simplified to:

$$D_i^K = 4850\, d_p \sqrt{\frac{T}{M_i}} \tag{1.8}$$

where d_p is the pore diameter in cm, T is the temperature in K, M_i is the molecular weight, and D_i^K is the diffusion coefficient in $cm^2 \cdot s^{-1}$.

1.2.5.4. Surface diffusion

When entering into even smaller-size pores with diameters comparable to the kinetic diameter of gas molecules, adsorption is no longer negligible. Recognized after the 1940s by Wicke-Kallenbach and Barrer, it was determined that the adsorption forces within pores of these sizes contributed clearly to the transport, and that such concept would explain the extra flow observed in experiments for microporous structures when analyzed with the Knudsen model. The diffusivities in this regime are also evidenced to follow an Arrhenius temperature dependence, for which it has been therefore referred to as an activated process. Surface diffusion, sometimes also called micropore diffusion or configurational diffusion, refers to the transport mechanism where the fluid molecules are not able to escape the force of the pore walls. It is usually encountered for d_p <2 nm and for strongly-adsorbed species also in somewhat larger pores [70, 72].

Diverse approaches for mixtures in the gas phase at the nanopore level considering this surface transport have been suggested. One of them is the approach by Krishna based on the Maxwell-Stefan model. This approach "relies on the empirical specification of the mutual-diffusion parameters from the individual transport coefficients [. . .] [70]." A detailed explanation of this model is presented in the next section.

Transport Mechanism	Definition	d_p size range	Selectivity	Mechanism's Depiction
Viscous flow/ molecular diffusion	d_p of the material is larger than the mean free path, being the molecule-molecule collisions much more frequent than the molecule-wall collisions.	> 50 nm	No separation $\alpha = 1$	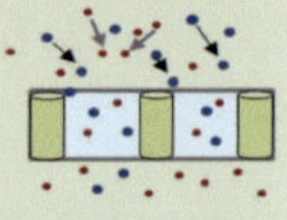 $d_p >> \lambda_m$
Knudsen Diffusion	Mean free path is larger than the d_p for which molecules collide more frequently with the walls than with other molecules.	2- 50 nm	$\alpha_{j/i} = \left[\frac{M_i}{M_j}\right]^{1/2}$	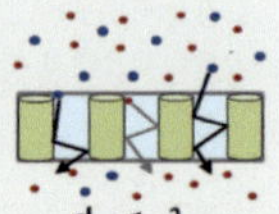 $d_p \leqslant \lambda_m$
Surface diffusion	As the d_p becomes even smaller, the molecules are not able to escape the force of the pore walls, i.e. adsorption forces, which must be also considered. *It also applies to larger pores with strongly adsorbed species.	1-3 nm*	Molecular sieving (absolute separation) Adsorption-diffusion (depending on ΔH)	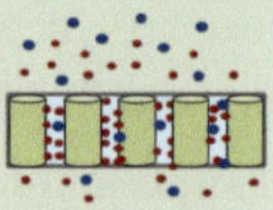 $d_p << \lambda_m$

Table 1.3: Diffusional transport mechanisms in porous materials [56, 63, 70, 72, 73, 74, 75]

1.2.6. Generalized Maxwell-Stefan surface diffusion model

Based on the Maxwell-Stefan equations, Krishna proposed in 1990 [76] the generalized Maxwell-Stefan surface diffusion model (GMS) seeking to describe the diffusion inside micropores under conditions where adsorption cannot be neglected considering the chemical potential gradient $\nabla\mu_i$ as the driving force [59, 70, 72]. When compared to a concentration gradient, the chemical potential gradient separates the kinetic and thermodynamic effects, permitting to take into account non-ideal behavior of mixtures [77].

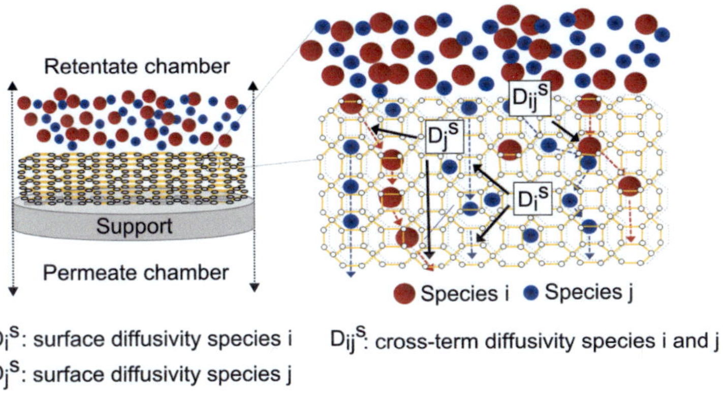

Figure 1.7: Maxwell-Stefan model depiction showing the surface diffusivities of components i and j (D_i^s, D_j^s) and the cross-term surface diffusivity (D_{ij}^s) [72]

Given the dominant surface forces in nanopores, the diffusing molecules enter an adsorbed state proceeded by a sequence of jumps. To aid in the comprehension of the model, Fig. 1.7 depicts molecules i and j, which are assumed to hop from one adsorption site to another. If a molecule i hops into a vacant site, an exchange of momentum between the site and the molecule occurs, in other words the friction experienced by molecule i from the vacant site (molecule-wall interactions); this is characterized by D_i^s, i.e., M-S surface diffusivity of component i also equivalent to the corrected diffusivity (apart from a geometric correction factor). If the site is rather occupied by a molecule of component j, the latter is displaced by component i, and a net change in the momentum of i and j takes place (molecule-molecule interactions in adsorbed state), in other words this refers to the facility of displacement of the adsorbed component j by component i. This is characterized by the D_{ij}^s, i.e., the cross-term surface diffusivity between component i and j in adsorbed state. Lastly, if the site is occupied by a molecule of the same species, this one is also displaced, but no net momentum change for the species will take place. Additional details on the diffusion coefficients can be found in further sections. The mathematical model is presented next [70].

1.2.6.1. Model

"The basis of the Maxwell-Stefan theory is that the driving force for movement that is acting on a species is balanced by the friction that is experienced by that species [78]." To represent the model, the equation below is introduced [78]:

$$-\frac{\theta_i}{RT}\nabla\mu_i = \sum_{\substack{j=1 \\ i\neq j}}^{n} \frac{\theta_j J_i^s - \theta_i J_j^s}{q_{sat}\cdot\rho_m\cdot D_{ij}^s} + \frac{J_i^s}{q_{sat}\cdot\rho_m\cdot D_i^s}, \qquad i,j = 1,2,...,n. \quad (1.9)$$

where θ_i and θ_j are the surface coverages of components i and j respectively, R is the universal gas constant, T is the temperature, $\nabla\mu_i$ is the chemical potential gradient, J_i^s and J_j^s are the fluxes of components i and j in the membrane layer respectively, q_{sat} is the saturation capacity, ρ_m is the density of the membrane, D_i^s is the M-S surface diffusivity of component i, and D_{ij}^s is the cross-term surface diffusivity between component i and j in adsorbed state.

In Eq. (1.9), the first term in the right side represents the friction resulting from the interaction between the adsorbed molecules, while the second term considers the interaction of the molecules of each component in adsorbed state with the pore wall, i.e., the driving force $\nabla\mu_i$, is the summation of the adsorbate-adsorbate interactions and the interactions between molecules and the pore wall for every component [78].

To explain a little deeper the chemical potential gradient and its relation to adsorption, this can be linked to the surface coverage by a matrix of thermodynamic factors, Γ_{ij}, as presented in the equation below [76].

$$\frac{\theta_i\nabla\mu_i}{R\cdot T} = \sum_{j=1}^{n}\Gamma_{ij}\nabla\theta_i \quad where \quad \Gamma_{ij} = \theta_i\frac{\partial\, lnp_i}{\partial\,\theta_j} \quad i,j = 1,2,...,n. \quad (1.10)$$

As shown, the surface coverage, θ_i, is related to the partial pressure by the adsorption model that determines then the form of the thermodynamic factor [78]. Some of the most known adsorption models include Langmuir, extended Langmuir, the ideal adsorbed solution theory (IAST), and the real adsorbed solution theory (RAST).

Langmuir The Langmuir model corresponds to the first description of adsorption on solid surfaces. This model is used in the hereby presented work to describe pure-component data. Its equation presented below, represents monolayer adsorption on a homogeneous adsorbent neglecting interactions between adsorbed molecules [79]:

$$\theta_i = \frac{q_i}{q_{sat}} = \frac{k_i \cdot p_i}{1 + k_i \cdot p_i} \tag{1.11}$$

where k_i is the adsorption constant of component i and p is the partial pressure of component i.

Knowing that $p_i = y_i \cdot P$, Eq. 1.11 can be rearranged to:

$$q_i = q_{sat}\Big(\frac{k_i \cdot y_i \cdot P}{1 + k_i \cdot y_i \cdot P}\Big). \tag{1.12}$$

Extended Langmuir One of the adsorption models chosen to be tested within the M-S surface diffusion model is extended Langmuir (eL). Besides its mathematical simplicity, this adsorption model has been proven to deliver adequate results for simple systems [78]. The surface coverage is described as:

$$\theta_i = \frac{q_i}{q_{sat}} = \frac{k_i \cdot p_i}{1 + \sum\limits_{j=1}^{n} k_j \cdot p_j} \qquad i, j = 1, 2, ..., n. \tag{1.13}$$

Combining Eq. (1.9) and Eq. (1.13), the flux of component i in a binary system can be obtained with:

$$J_i^s = -\frac{q_{sat} \cdot \rho \cdot D_i^s}{1 - \theta_i - \theta_j} \cdot \frac{(1 - \theta_j + \theta_i \frac{D_j^s}{D_{ij}^s})\nabla\theta_i + (\theta_i + \theta_i \frac{D_j^s}{D_{ij}^s})\nabla\theta_j}{\theta_j \frac{D_i^s}{D_{ij}^s} + \theta_i \frac{D_j^s}{D_{ij}^s} + 1}. \tag{1.14}$$

Ideal adsorbed solution theory (IAST) Another model chosen to be studied in this work corresponds to the IAST, firstly introduced in 1965 by Myers and

Prausnitz [80] becoming quickly popular within the adsorption science community. This approach is frequently used to validate and evaluate experimental data as well as new theories for adsorption equilibria of gas mixtures [81]. This model has been generally proven to deliver good simulation results in the following scenarios: 1) close to ideal mixtures and 2) components with similar adsorption loadings, while posing challenges on 1) mixtures of polar species [82].

One of the criteria of an ideal adsorbed solution is the lack of interaction between species [82]. The thermodynamic factor, with no analytic solution, can then be described as:

$$\Gamma_{ij} = \theta_i \frac{\partial\, lnp_i}{\partial\, \theta_j} = \begin{cases} \frac{\theta_i}{p_i}\frac{\partial p_i}{\partial \theta_i} & j = i \\ 0 & j \neq i \end{cases} \tag{1.15}$$

Based on single-component isotherms, the pressure in the gas phase and the spreading pressure of the adsorbed components on the surface should be equal. Additionally, IAST is analogous to Raoult's law for vapor-liquid equilibrium [82]:

$$p_i = y_i \cdot P = x_i \cdot p_i^0(\pi) \tag{1.16}$$

where $p_i^0(\pi)$ is the partial pressure of single component i calculated at the temperature and spreading pressure of the mixture. Based on these, the molar fractions of species i, x_i, in all adsorbed molecules on the surface may be determined by solving a system of n nonlinear equations [59]:

$$\begin{bmatrix} \sum_{i=1}^{n} x_i - 1 \\ \vdots \\ q_{sat,i}\, ln(1 + k_i \frac{p_i}{x_i}) - q_{sat,i+1}\, ln(1 + k_{i+1} \frac{p_{i+1}}{x_{i+1}}) \\ \vdots \\ q_{sat,n-1}\, ln(1 + k_{n-1} \frac{p_{n-1}}{x_{n-1}}) - q_{sat,n}\, ln(1 + k_n \frac{p_n}{x_n}) \end{bmatrix} = \mathbf{0} \tag{1.17}$$

To calculate then the coverage of species, the following equation is used [59, 82]:

$$\theta_i = \frac{q_i}{q_{sat,i}} = \frac{q_{tot} \cdot x_i}{q_{sat_i}} = \frac{x_i}{q_{sat,i} \sum\limits_{i=1}^{n} \frac{x_i}{q_i^0}}, \tag{1.18}$$

where q_{tot} is the total loading of adsorbed molecules and q_i^0 is the standard state loading, i.e., the amount of component i adsorbed at constant temperature and spreading pressure in absence of other species.

Using the Langmuir model for the pure-component data, Eq. (1.12) can be then substituted in Eq. (1.18) for components i and j:

$$\theta_i = \frac{x_i}{q_{sat,i}} \left[\frac{x_i \cdot (1 + k_i \cdot y_i \cdot P)}{q_{sat,i} \cdot k_i \cdot y_i \cdot P} + \frac{x_j \cdot (1 + k_j \cdot y_j \cdot P)}{q_{sat,j} \cdot k_j \cdot y_j \cdot P} \right]^{-1}. \tag{1.19}$$

Real adsorbed diffusion theory (RAST) Derived from the IAST, the real adsorbed diffusion theory is another adsorption model that takes into account deviations from ideality usually attributed to surface energetic heterogeneity of the sorbent or activity coefficients $\neq$1. The model relies on multicomponent experimental data in order to estimate the activity coefficients, i.e., it is not based on prediction, rather on experimentation [83]. Furthermore, another version of this model called the predictive real adsorbed solution theory (PRAST) does use pure-component isotherm data to obtain the sorbate activity coefficients [84].

1.2.6.2. Diffusivities

M-S surface diffusivity D_i^s refers to the M-S surface diffusivity of component i and apart from a geometric correction factor has been found to be equal to the corrected diffusivity. The latter is obtained via the Darken equation [85] and is said to be a correction for the Fickian diffusivity based on stating the concentration gradient as the driving force for diffusion [78]. For a unary system, the M-S surface diffusivity D_i^s can be calculated as shown in Eq. (1.20). Single gas permeation experiments must be conducted and adsorption parameters must also be known, or likewise determined through experimentation [59, 78].

$$D_i^s = -\frac{J_i^s}{\rho_m \cdot q_{sat} \cdot \nabla\theta_i} \cdot (1 - \overline{\theta_i}) \tag{1.20}$$

where ρ_m is the density of the material, $\nabla\theta_i$ is the gradient of the coverage of component i along the membrane, and $\overline{\theta_i}$ is the average coverage of component i in the membrane.

Cross-term surface diffusivity D_{ij}^s represents the cross-term surface diffusivity, also referred to as counter-exchange diffusivity, between component i and j, i.e., the diffusivity that accounts for the facility of replacement of component i at an adsorbed site by component j [77]. In order to estimate D_{ij}^s, the self-exchange diffusivity $D_{i,i}^s$ must be determined [59]:

$$D_{i,i}^s = \theta_i \cdot \left[\frac{1}{D_{self,i}^s} - \frac{1}{D_i^s}\right]^{-1} \tag{1.21}$$

where $D_{i,i}^s$ represent the M-S self-exchange diffusivity and $D_{self,i}^s$ the self-diffusivity of component i. Nevertheless, the assessment of the latter poses a great challenge due to the requirement of demanding experimental methods such as nuclear magnetic resonance pulsed field gradient spectroscopy (PFG-NMR) or quasielastic neutron scattering (QENS) or equally demanding molecular dynamics simulations (MD) [59]. Given such complex task, the estimation of D_{ij}^s as a function of the M-S surface diffusivities was proposed by Krishna [76] as a generalized form of the Vignes empirical correlation [78]:

$$D_{ij}^s = {D_i^s}^{\frac{\theta_i}{\theta_i + \theta_j}} \, {D_j^s}^{\frac{\theta_j}{\theta_i + \theta_j}} \tag{1.22}$$

which also satisfies the Onsager reciprocity relation [76]:

$$D_{ij}^s = D_{ji}^s. \tag{1.23}$$

1.3. Tasks and objectives

ZIF-8 is characterized by pore apertures of 3.4 Å, which range within the kinetic diameters of several gas molecules [38]. Various works have been published involving the highly successful separation of hydrocarbons [13, 86, 87, 88, 89, 90], but ethene/ethane separation remains still a bigger challenge being the highest reported separation factor until now as low as 2.8 [91]. This research work concentrates in the synthesis of ZIF-8 SURMOF membranes using gold-coated alpha alumina supports via LPE LBL synthesis for application in gas separation of binary (alkane/alkene) gas mixtures, in this case ethene/ethane separation. Likewise, the work is complemented with a simulation study using the generalized Maxwell-Stefan surface diffusion model (GMS) proposed by Krishna in 1990 [76], seeking to analyze the diffusional transport within the framework in a quantitative way. The main objective is to be able to assess the performance of the LPE LBL deposition via the dipping technique on gold-coated porous alpha alumina supports for application in alkane/alkene separation to determine the potential of such technique and material in further studies, e.g., tailoring of SURMOF with alternative linker groups, as well as to analyze the experimental separation performance of the synthesized membrane with diffusion theory via a simulation program.

To begin, **Chapter 2: ZIF-8 SURMOF membrane synthesis and characterization** concentrates in the synthesis results of the membranes. Initially, two supports are studied. Data of the characterized membranes including XRD, IRRAS, and SEM is presented. Focusing then on one of the supports, H_2 pressure-dependent experiments are also shown in order to help determine the optimal number of deposition cycles required to obtain an overall high membrane quality.

Chapter 3: Thermal and mechanical stability of ZIF-8 SURMOF membranes, seeks to define the temperature and pressure limits of the studied system. The gas separation experiments were performed at room temperature and absolute pressures ranging from 1.1 to 1.3 bar, for which proving the stability of the system under these conditions is of primary concern to be able to confidently use the experimental data obtained later in the simulation study.

Chapter 4: ZIF-8 SURMOF membrane-based alkane/alkene gas separation focuses on the membrane performance in ethene/ethane separation. Different scenarios are studied and analyzed; separation factors, as well as permeances are reported and compared with other published literature. The counter-flux of sweep gas into the retentate chamber is also revised.

To finish, **Chapter 5: Simulation model adapted for a ZIF-8 SURMOF system** studies the relationship of the obtained experimental data with diffusion theory. The simulation study is based on the previously presented generalized Maxwell Stefan surface diffusion model, with a chemical potential gradient described via two different adsorption models: extended Langmuir and IAST. The goal of the simulation is to determine the adsorbed and molar fractions of both components within the thin film as well as the components' fluxes, based on a discretization of the membrane that could help understand the implications of the diffusion mechanism through the framework.

All of the chapters begin by stating their objective, followed then by experimental procedures (if applicable), proceeding then with the presentation of results. A summary, as well as an outlook of the presented work is also included as **Chapter 6**. Further information referenced within the chapters can also be found in the appendixes.

This study was funded by the Helmholtz Research School Energy-Related Catalysis.

2. ZIF-8 SURMOF membrane synthesis and characterization

Given the remarkable thermal and mechanical stability that ZIF materials possess, intense research has been dedicated towards their application within the membrane technology field over the last 18 years [11]. Diverse synthesis approaches including direct synthesis [91, 92, 93, 94], secondary growth [38, 95], surface special seeding [96, 97], among others have been thoroughly explored. Still, the achievement of reproducible defect-free highly-oriented monolithic and high performing ZIF-8 SURMOF membranes remains a challenge. Therefore, approaches targeting such issues continue to be proposed and investigated.

Epitaxially grown MOFs have proven to be a promising approach consisting in the use of interfaces to initiate and control their growth in terms of orientation and crystallinity [24]. The liquid phase epitaxial (LPE) layer-by-layer (LBL) approach was systematically studied on two different supports. First, a comparative study of both approaches based on characterization techniques like XRD, IRRAS, and SEM was performed. Furthermore when speaking about membranes, achieving highly oriented crystals is not the only factor to be deeply considered. Producing surfaces without pinholes, cracks, and/or defects is of high importance in order to minimize the non-selective gas transport and have a true chance in gas separation. Thus, the number of deposition cycles required to fully cover the support's surface was also studied by determining the permeance of H_2 as a function of increasing pressure. Although this technique helps to determine whether large defects exist within the membranes, other types of micro-defects that could still affect the selectivity of the membranes cannot be determined with this procedure. This behavior was later on studied and is reported in Chapter 4. The information presented in this chapter was already published in [35] though more details are included in the present version.

2.1. Substrate description and preparation

In this work, porous α-Al_2O_3 membranes (Fraunhofer IKTS; d=13 mm, h=1 mm, support d_{50}=2.5 μm, membrane d_{50}=70 nm, ε_s=0.4-0.55) were used as support and further divided into two groups: non-coated and Au-coated supports. Non-coated supports were activated via UV-light exposure during 1 hour using the UV/Ozone Procleaner^{TM} (240 V AC, 50/60 Hz, 0.3 A, 1 PH) from Bioforce Nanosciences for the functionalization of the surface support with -OH groups. The other group of supports was first covered with a thin film of gold ($\approx$ 70 nm) via physical vapor deposition (PVD). This procedure was performed by indirect resistance heating in which gold is evaporated through an energy source at a defined temperature. The substrate was placed inside a vacuum chamber and heated at around 100 °C ensuring a good coating adhesion. Once coated and before proceeding with the preparation of these supports, SEM images and further tests were conducted to ensure the thin Au film applied was not compromising the porosity of the support, i.e., blocking the pores [35].

First, cross section and surface SEM images were made to exclude pore blockage by the gold coating as shown in Fig. 2.1. A clear increase of brightness in the gold-coated porous α-Al_2O_3 SEM images can be observed; this as a result of a stronger backscattering of electrons due to the presence of gold atoms on the support. Nonetheless, the pore layers are still completely visible and no distinct major particle conglomeration is observed, i.e., no obvious pore blockage can be observed. The cross section images also show a granular structure for both cases with no observable diffusion of the deposited gold along the cross section of the support [35].

However, in order to confirm the latter point an electron probe microanalysis was conducted and the results are shown in Fig. 2.2. The analysis was done for elements a) O, b) Al, and c) Au confirming that the first two are concentrated in the main volume of the substrate, while the gold is localized exclusively in the top layer. The microprobe analysis for the cross section elemental analysis was obtained with a JXA-8530F from Jeol, Japan with a 20 kV beam at 11 mm working distance and a dwell time of 20 ms. Additionally, permeation experiments were also conducted on the non-coated and Au-coated substrates, further supporting no obvious pore blockage which would hinder then the diffusion during gas separation experiments. Variations among the supports were also studied via permeance vs. pressure experiments [35]. More detailed information regarding this point can be found in Appendix A.2.

Once it was confirmed that the porosity of the supports was not compromised, an -OH functionalization on the Au coating was achieved by immersing it in a 1 mM ethanolic solution of 11-mercapto-1-undecanol (MUD, Sigma Aldrich) for 24 hours. Afterwards, the substrate was rinsed with ethanol and dried with N_2.

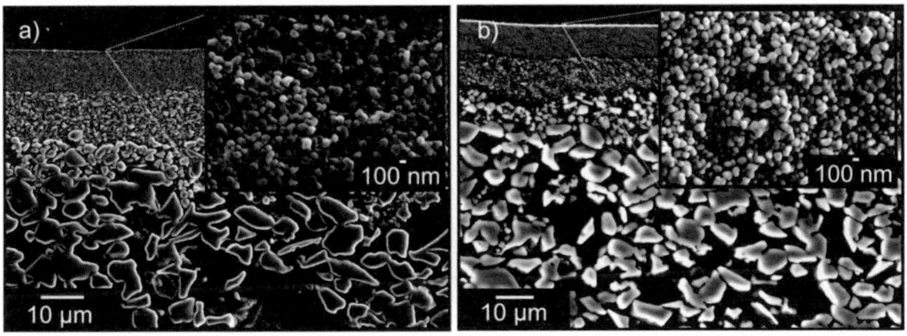

Figure 2.1: Surface and cross section images of the a) non-coated and b) Au-coated α-Al_2O_3 supports (adapted from [35])

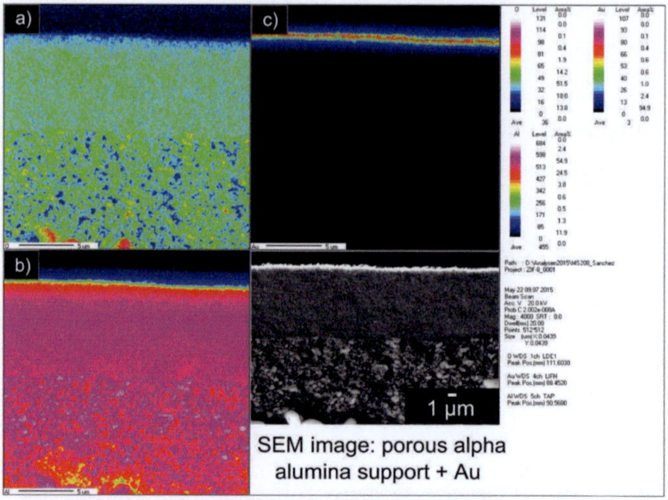

Figure 2.2: Cross section element mapping of an Au-coated α-Al_2O_3 support: a) Oxygen, b) Aluminum, c) Gold. (adapted from [35])

2.2. ZIF-8 SURMOF preparation

For all of the synthesis experiments regardless of the support, the following procedure was performed. Using a dipping robot operating at normal temperature and pressure, the substrate was sequentially immersed in the methanolic and rinsing solutions in the following order for defined times:

1. Metal source: 10 mM zinc dinitrate hexahydrate 98% (Alfa Aesar) solution for 300 s
2. Rinsing step: pure methanol (Merck seccosolv) for 100 s
3. Linker source: 20 mM 2-methyl-1H-imidazole (Merck) for 300 s
4. Rinsing step: pure methanol (Merck seccosolv) for 100 s

After step 4, a cycle is finished and the next cycle starts over again from step 1. According to literature [24], the number of cycles permits a direct control over the thickness of the SURMOF. After synthesis completion, the substrate was rinsed and dried with N_2. This work studied different numbers of deposition cycles scenarios including 100, 125, 150, 170, 175, and 200 [35].

2.3. Techniques used for ZIF-8 characterization

X-ray diffraction The XRD out-of-plane data presented in this work was obtained applying Cu-K$\alpha_{1,2}$ radiation of 0.15419 nm using a Bruker D8 Advance diffractometer equipped with a Si-strip detector (PSD Lynxeye©) in $\theta - \theta$ geometry and a variable divergence slit on the primary side. The diffractograms were recorded over an angular range of 2θ= 5.5-20° and 1) 84 s or 2) 168 s per step.

Infrared reflection absorption spectroscopy IRRA spectra were performed using a Vertex 80 FT-IR spectrometer from Bruker Optics equipped with a grazing incidence sample accessory providing an angle of 80° relative to the surface normal and a liquid nitrogen-cooled mercury cadmium telluride (MCT) narrow band detector. The measurements were done at ambient conditions. For the background measurements, perdeuterated hexadecanethiol SAMs (synthesized and provided by the A. Terfort group in Frankfurt) on gold-coated α-Al_2O_3 porous supports were used.

Scanning electron microscopy Electron microscopy images were performed with the following devices: an XL30 ESEM-FEG with a 20 kV beam in SE mode at 10 mm working distance, a JSM-6300 from Jeol, Japan with a 10 kV beam in SE mode at 8 mm working distance, and a JXA-8530F from Jeol, Japan with a 20 kV beam in SE mode at 11 mm working distance. Samples were sputter-coated with 10 nm of gold/palladium.

2.4. Comparative growth study on various porous α-Al_2O_3 supports

2.4.1. Synthesis results and discussion

To facilitate the comprehension of this section, the results for the supports without a gold coating (non-coated) are presented in blue color, while the results for the Au-coated supports are presented in red color [35].

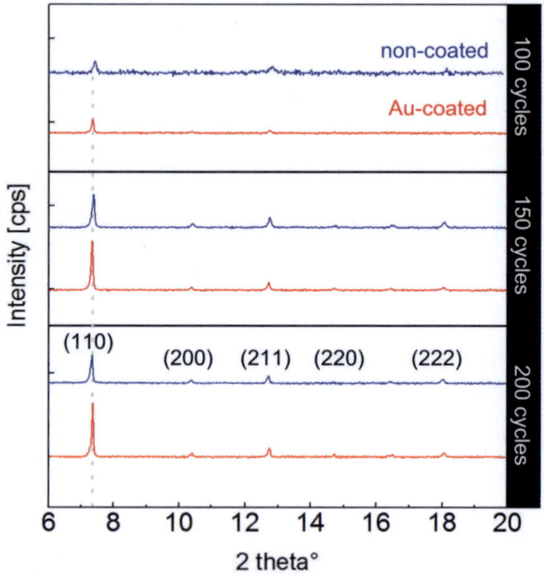

Figure 2.3: XRD patterns of ZIF-8 SURMOFs deposited on non-coated and Au-coated α-Al_2O_3 supports for 100, 150, and 200 deposition cycles (adapted from [35])

To begin, ZIF-8 SURMOF LPE LBL dipping experiments were conducted on both types of supports for different numbers of deposition cycles: 100, 150, and 200. The XRD results for the non-coated and Au-coated supports are shown on

Fig. 2.3. As it can be observed, all of the scenarios resulted in ZIF-8 SURMOF growth with a predominant orientation in the (110) plane, besides other oriented components expected due to the porosity of the supports. Nonetheless, in all of the Au-coated cases, the relative intensity of the characteristic (110) reflex at $2\theta = 7.36°$ was higher and increased in general with higher numbers of deposition cycles compared to those of the non-coated samples [35]. In order to evaluate the growth orientations on both supports more easily, Fig. 2.4 shows overlapping normalized diffractograms for the 200 deposition cycles scenarios, as well as the powder diffractogram of ZIF-8; the latter one was extracted from Lewis's published work [46]. The diffractrograms were normalized to the highest peak intensity, in this case the (110) reflex. It could be concluded that the ZIF-8 SURMOFs grown on both supports show a decrease in peak intensity for the other planes while an increase in intensity can be observed for the (110) plane. This reduction was much more pronounced for the Au-coated supports reaching a 56% and 59% decrease on the (211) and (222) plane reflexes, respectively; the non-coated supports presented an intensity reduction of 21% and 26% in the (211) and (222) planes, respectively.

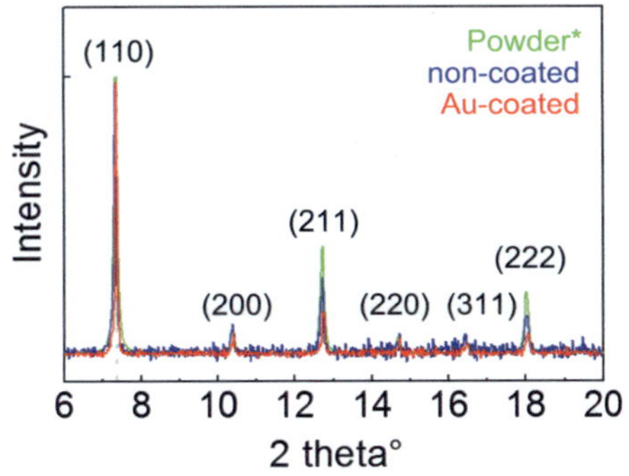

Figure 2.4: Normalized XRD patterns of ZIF-8 powder [46] and ZIF-8 SURMOFs deposited on non- and Au-coated α-Al_2O_3 supports (200 cycles)

Additionally, the integral breadths of the peaks were also obtained using the Diffrac.EVA V2.1 program from Bruker with a Scherer constant of k_{sch}=0.89. The results reported lower values and hence bigger domain sizes for the crystals on the (110) and (220) planes, further supporting a predominant orientation in the [110]-direction. The estimated crystal size domain corresponded to $\approx$ 350 nm and $\approx$ 170 nm for the Au-coated and non-coated supports, respectively. Tables regarding these information can be found in Appendix A.3.

To further support this information, SEM images shown in Fig. 2.5 were obtained. The a) non-coated supports confirmed the synthesis of ZIF-8 crystals, with different sizes, orientations, and common agglomerates. Additionally, there were cases in which the support presented large uncovered areas, which would definitely hinder gas separation tests. In contrast to the non-coated substrates, for the b) Au-coated supports the deposition was much more homogeneous in terms of size and roughness when compared to the non-coated supports. With an increasing number of deposition cycles higher density of deposited material was clearly observable as well as the formation of some small agglomerates [35].

Moreover, the surface composition of the ZIF-8 SURMOF membranes on the Au-coated supports was also studied by IRRAS. As it can be observed in Fig. 2.6, all of the cases above reported the characteristic IR bands for ZIF-8, including the C-N absorption bands between 1100 and 1400 cm^{-1}, the methyl bending mode at 1384 cm^{-1}, as well as the mode at 1584 cm^{-1} associated to the C=N stretch, and the 2929 cm^{-1} and 3135 cm^{-1} modes assigned to the aromatic and aliphatic C-H stretching of the imidazole, respectively. Additionally, all of the obtained spectra were consistent confirming the reproducibility of the ZIF-8 synthesis without any unexpected compounds [35].

Based on these results, it is noticeable that under the same experimental conditions the Au film on the α-Al_2O_3 porous supports – compared to the non-coated supports – aids in obtaining higher oriented deposited crystals on the support's surface. Some of the possible reasons improving these properties could be related to a more efficient anchoring of the SURMOF on the Au surface by the immersion of the support in the MUD solution before SURMOF growth, causing the formation of a thiol SAM, which activates the surface for SURMOF growth. In addition, a slight reduction of the surface roughness when applying the thin Au coating could further influence the MOF crystallites' orientation and therefore their properties [35].

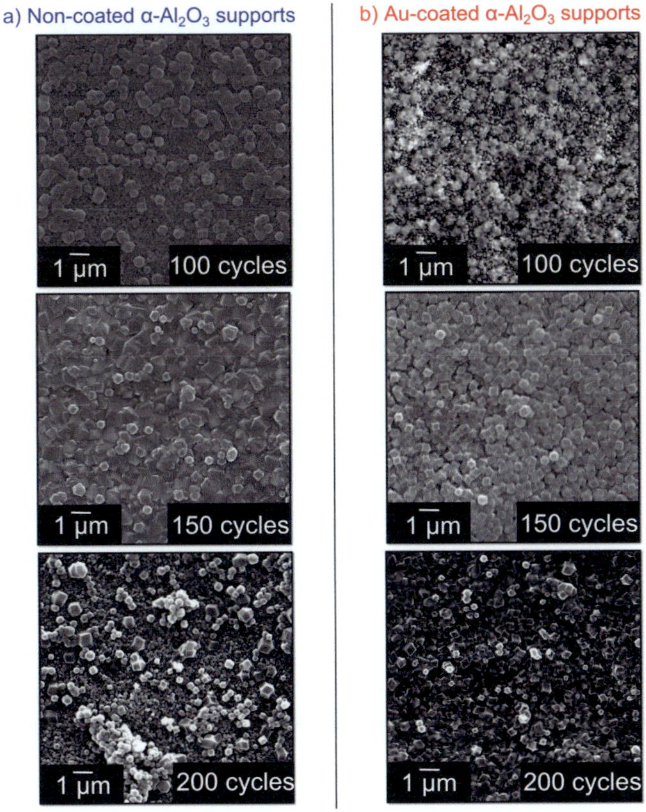

Figure 2.5: SEM images of ZIF-8 SURMOF deposition on a) non-coated and b) Au-coated α-Al_2O_3 supports for various deposition cycles (adapted from [35])

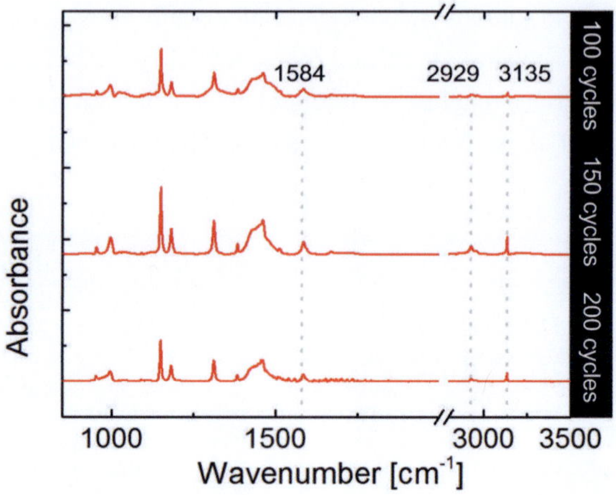

Figure 2.6: IRRA spectra of ZIF-8 SURMOFs deposited on Au-coated α-Al_2O_3 supports for 100, 150, and 200 deposition cycles (adapted from [35])

2.5. Pressure-dependent experiments

To study the ZIF-8 SURMOF deposition process and with it the membranes' overall quality, pressure-dependent permeance experiments using H_2 as the feed gas at increasing pressures were performed.

2.5.1. Experimental methodology

The membranes were mounted inside 4-end modules consisting of two chambers possessing two inlets and two outlets. Fig. 2.7 shows a depiction of this 3D-printed modules designed at IMVT and printed at IFG with an Eden 260V printer from Stratasys using VeroWhite and VeroClear PolyJet materials; their design ensures an even gas flow all over the surface of the membrane. The membranes were glued in the lower chamber (facing the SURMOF upside) and sealed gastight to the upper chamber using ELASTOSIL® M 4670 A/B from Wacker Chemie AG.

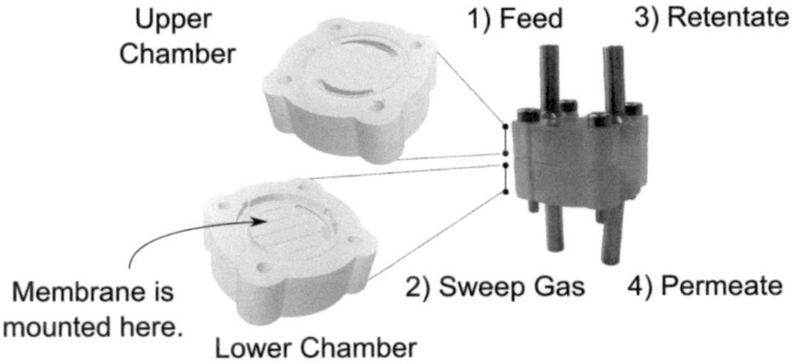

Figure 2.7: 4-end 3D-printed membrane module composed of an upper and a lower chamber possessing two inlets: 1) feed, 2) sweep gas, and two outlets: 3) retentate flow and 4) permeate flow

Once mounted inside, the membranes were tested using H_2 (150 ml·min^{-1}) as the inlet flow at different overpressures (1.16 -1.41 bar) and Ar (150 ml·min^{-1}) as the sweep gas set to an initial P_{abs}= 1.1 bar; the retentate and permeate flows were then recorded using flow meters.

2.5.2. Permeance results and discussion

The next step in the characterization of the ZIF-8 SURMOF membranes consisted in verifying the existence or absence of cracks and/or large defects. To explain this, it is important to understand that the different possible transport mechanisms through the pores, as explained in Chapter 1, depend on the pore size.

One of the parameters usually investigated for transport mechanisms is the pressure and permeance dependency. Fig. 2.8 shows schematically the behavior of permeance with increasing pressures for viscous flow, Knudsen diffusion, and surface diffusion [63]. As it can be seen, in the Knudsen and surface diffusion regimes an increase in pressure should not affect significantly the permeance behavior. On the other hand, a considerable increase in permeance is characteristic of viscous flow. Due to the ZIF-8's pore window diameter of about 3.4 Å, almost constant permeance values are expected to be measured with increasing pressures.

If an increase in permeance were to be measured, this would imply the presence of cracks and/or large defects, with sizes larger than 50 nm such as uncovered support areas.

Furthermore, it is important to consider that – regardless of being in the surface diffusion regime – a decrease in permeance is possible with increasing feed pressure when the surface concentration on the feed side climbs beyond the linear part of the adsorption isotherm, i.e., when the transport limit has been reached [35].

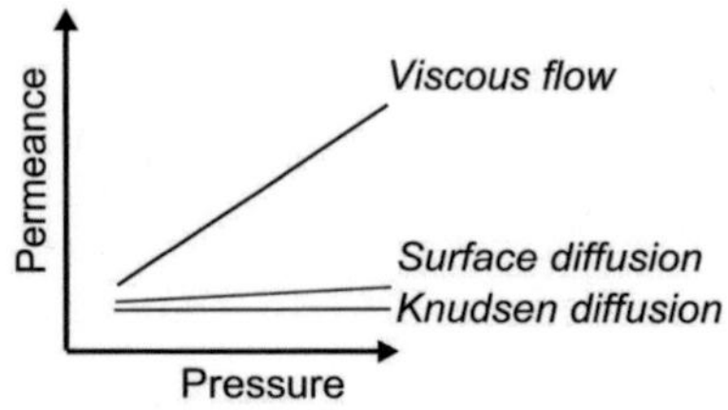

Figure 2.8: Permeance profiles of different transport mechanisms for increasing pressures [63]

To study this point, transmembrane pressure-dependent experiments were performed on the ZIF-8 SURMOF membranes. The permeance [$mol{\cdot}s^{-1}{\cdot}m^{-2}{\cdot}Pa^{-1}$] was calculated with the following equation [98]:

$$Permeance = \frac{Permeation flux}{Partial\ pressure\ difference}. \tag{2.1}$$

In Fig. 2.9 the permeance as a function of the H_2 pressure for different numbers of deposition cycles is shown. The graph shows the results of three ZIF-8 SURMOF membranes on Au-coated α-Al_2O_3 supports (100, 125, and 150 deposition cycles) and one ZIF-8 SURMOF membrane on a non-coated α-Al_2O_3 support (150 deposition cycles). As it can be observed, the experiments ran for 100 and 125 cycles still showed the characteristic permeance profile of viscous flow probably due to partially uncovered areas of the support. Nonetheless, when testing the sample prepared with 150 deposition cycles, an almost constant permeance behavior could be observed, suggesting that 150 cycles successfully achieves a fully covered support. When performing the same test for the support without Au coating constant permeance was not achieved, not even within the 150 cycles scenario. This could possibly be related to defects or non-coated zones on the substrate as supported by previous SEM images (see Fig. 2.5). For these and the previous analysis already discussed in this chapter, the Au-coated supports were selected for all further studies in this work [35].

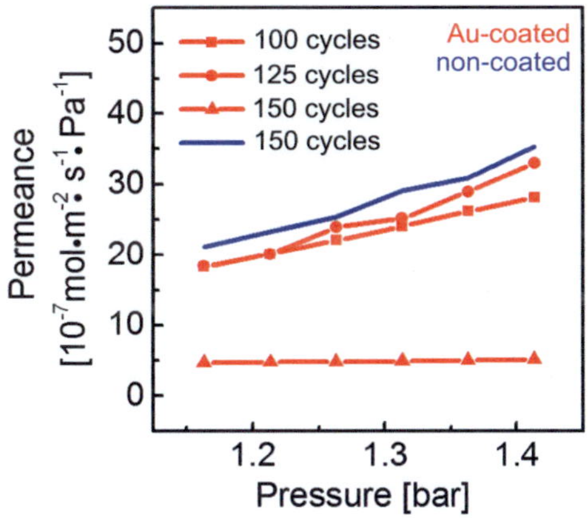

Figure 2.9: H_2 permeance through ZIF-8 SURMOF LPE LBL membranes with different numbers of deposition cycles tested for increasing feed absolute pressures (1.16 -1.41 bar) at room temperature (adapted from [35])

To corroborate the least number of deposition cycles required to completely cover the Au-coated α-Al_2O_3 supports as well as the reproducibility of the technique, additional 150 deposition cycles experiments were conducted. As shown in Fig. 2.10, samples 2, 3, and 4 resulted in comparable profile results. The permeance values obtained varied between 0.8 and 4.8 $\cdot$ 10^{-7}mol$\cdot$m^{-2}$\cdot$s^{-1}$\cdot$Pa^{-1} comparable to similar works using other synthesis techniques published by McCarthy, Pan, and Bux [93, 99, 100, 101]. Nonetheless, sample 1 in Fig. 2.10 presented a profile definitely corresponding to viscous flow. Fig. 2.11 shows the SEM image of a membrane with similar permeance behavior to sample 1. It is obvious that though the synthesis of well-defined and homogeneously sized cubic crystal structures was successful within 150 deposition cycles, still some non-coated areas of the α-Al_2O_3 support exist [35]. As mentioned at the beginning of the chapter, the α-Al_2O_3 supports possess a membrane surface layer of d_{50}= 70 nm. Nonetheless, when studying the supports it was proven that no two α-Al_2O_3 supports are exactly the same for which such growth variations are believed to be associated in part to the supports themselves; more information on this can be found in Appendix A.2. Due to these support variations, it might be possible that 150 deposition cycles do not always guarantee that the ZIF-8 SURMOF fully covers the Au-coated α-Al_2O_3 supports.

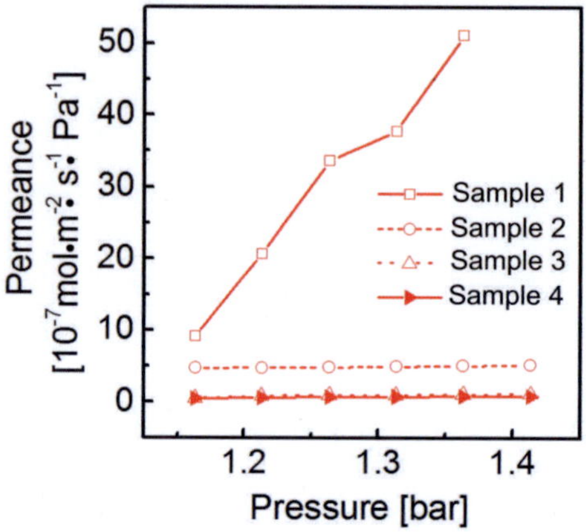

Figure 2.10: H_2 permeance through ZIF-8 SURMOF LPE LBL membranes with 150 deposition cycles tested for increasing feed absolute pressures (1.16 -1.41 bar) at room temperature (adapted from [35])

To continue exploring the number of deposition cycles required, further experiments were conducted for 170 and 175 cycles. All of the cases resulted once again in defined crystallinity without traces of unexpected compounds as shown in Fig. 2.12 a) and b), respectively. To facilitate the comprehension of this and following sections, the samples are color-coded and will remain so all over the work. After confirming their crystallinity and chemical composition by XRD and

IRRAS respectively, H_2 pressure-dependent experiments were performed. The results are reported in Fig. 2.13. All of the samples show the expected profile with H_2 permeance values between 1.27 and 26.65 $\cdot 10^{-7}$mol$\cdot$m$^{-2}\cdot$s$^{-1}\cdot$Pa^{-1}. These data are comparable to published literature working with other synthesis techniques reporting values between 0.6 and 15.4 $\cdot 10^{-7}$mol$\cdot$m$^{-2}\cdot$s$^{-1}\cdot$Pa^{-1} [93, 99, 100, 101]. Based on the profiles obtained and the reproducibility of the results, it can be defined that 170 is the minimum number of deposition cycles required to achieve reproducible monolithic ZIF-8 SURMOF membranes via the LPE LBL deposition technique at normal temperature and pressure. Nonetheless, it is interesting to mention that all of the samples presented different permeances which should play an important role in gas separation experiments. For instance, sample 1 and 2 probably possess more grain boundary micro-defects not detectable via pressure-dependent experiments, but that will definitely play a role in the performance of the membrane. On the other hand, samples 3 and 4 present lower permeances and hence should possess less micro-defects [11, 13] which should result in higher separation efficiencies and increased selectivities.

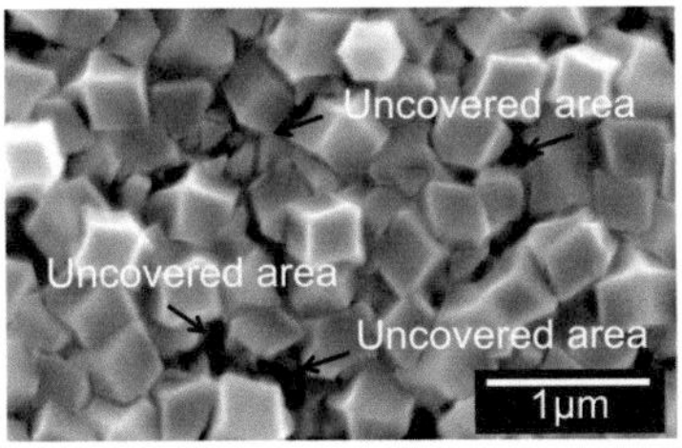

Figure 2.11: SEM image of a ZIF-8 SURMOF LPE LBL membrane with 150 deposition cycles presenting uncovered support areas (adapted from [35])

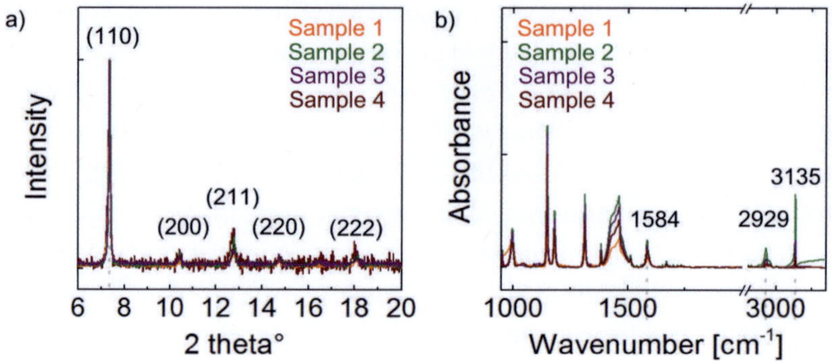

Figure 2.12: a) Normalized XRD patterns and b) IRRA spectra of ZIF-8 SURMOFs deposited on Au-coated α-Al_2O_3 supports for 170 (Sample 4) and 175 (Samples 1-3) deposition cycles

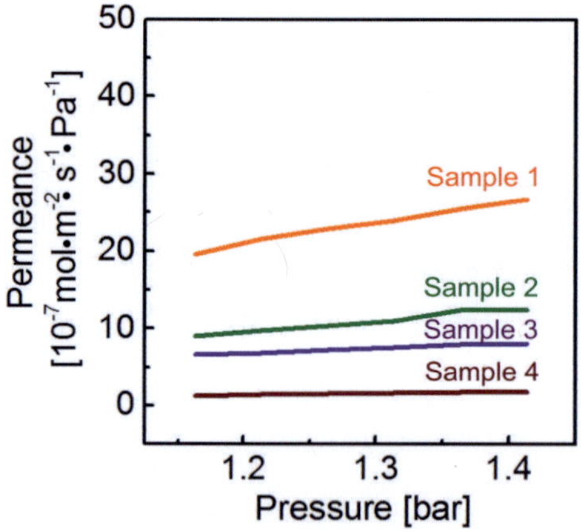

Figure 2.13: H_2 permeance through ZIF-8 SURMOF LPE LBL membranes with 170 (sample 4) and 175 (samples 1-3) deposition cycles tested for increasing feed absolute pressures (1.16 -1.41 bar) at room temperature

3. Thermal and mechanical stability of ZIF-8 SURMOF membranes

ZIF-8 is one of the most studied MOFs, primarily due to its recognized thermal stability. It has been proven to withstand temperatures of up to 550 °C in N_2 and up to 300 °C in the presence of water or oxidative environments [19, 102]. Nonetheless when speaking about thin films or membranes, the behavior of this material is expected to vary for instance due to thermomechanically induced stresses between the crystals and the support [52]. Up to now, data regarding ZIF-8 temperature stability in thin films is limited and might differ from system to system, being of high importance in order to determine the suitability of the material for diverse applications, e.g., gas separation, catalysis, among others.

Moreover, simulated data does not always agree with experimental data. If this is the case, such disagreement should not be instantly attributed to a misfit in the model, but rather also questioned whether certain unconsidered criteria (e.g., system's stability, measurement errors) might have been influencing the experimental data obtained.

In order to cover such points, this chapter sought to define the temperature and pressure limit of the LPE LBL ZIF-8 SURMOF established system. Due to the fact that the membranes are meant to be tested for gas separation at normal temperature and pressure, the crucial point was to define the stability of the crystals under ethene and ethane exposure at such conditions. The experimental results obtained in this chapter were then taken as a reliable base for the simulation runs further discussed in Chapter 5.

3.1. Techniques used for ZIF-8 characterization

As-synthesized, treated (exposed to a constant stream of ethene or ethane for 24 hours at room temperature, 100 °C, 150 °C, or 200 °C), and mechanically compressed (6.5, 50, or 100 N) membranes were characterized with diverse techniques which are presented next.

X-ray diffraction XRD out-of-plane patterns were obtained applying Cu-K$\alpha_{1,2}$ radiation of 0.15419 nm using a Bruker D8 Advance diffractometer equipped with a Si-strip detector (PSD Lynxeye©) in $\theta - \theta$ geometry and a variable divergence slit on the primary side. The diffractograms were recorded over an angular range of 2θ= 5.5-20° and 168 s per step.

Infrared reflection absorption spectroscopy IRRA spectra measurements were performed using a Vertex 80 FT-IR spectrometer from Bruker Optics equipped with a grazing incidence sample accessory providing an angle of 80° relative to the surface normal and a liquid nitrogen-cooled mercury cadmium telluride (MCT) narrow band detector. The measurements were done at ambient conditions and perdeuterated hexadecanethiol SAMs on gold-coated α-Al_2O_3 porous supports were used for the background measurements.

Scanning electron microscopy Electron microscopy images were performed using a Zeiss Leo 1530 with a 5 kV EHT and 6.3-6.7 mm working distance; the samples were sputter-coated with a 10 nm thick gold layer.

3.2. Stability treatments

3.2.1. Experimental methodology

A systematic approach was followed in order to investigate the effect of temperature on ZIF-8 crystals under exposure of ethene or ethane. As a first step, the Au-coated α-Al_2O_3 supports were broken into two pieces. Following, ZIF-8 SURMOF membranes were synthesized with 175 deposition cycles following the procedure previously described in Chapter 2. After synthesis and prior to any other treatment, i.e., as-synthesized, the membranes were characterized with XRD and IRRAS. As a next step, one piece of each membrane was directly used for SEM studies in order to investigate the shape and size of the as-synthesized ZIF-8 crystals. The second piece of the sample was then treated in a Heraeus tubular oven type Ro 4/25 with a maximum operation temperature of 1110 °C for 24 hours under a constant stream of ethene or ethane at diverse temperatures: room temperature, 100 °C, 150 °C, or 200 °C. Then, the treated membranes were characterized with XRD and IRRAS in order to determine any change in crystallinity or chemical composition, as well as with SEM to obtain a visual comparison of the ZIF-8 crystals prior and after treatment.

3.2.2. Results and discussion

3.2.2.1. Room temperature

The first and most important point in this chapter was to confirm the ZIF-8 membrane stability at room temperature. After running the experiments, no significant changes in crystallinity were observable by XRD under ethene or ethane exposure as shown in Fig. 3.1. Furthermore, using the Williamson and Hall model, which does not only consider size, but also stress and strain broadening, the mean scattering domains of the crystals were determined [103]. It is important to mention that this method proves to be helpful when used in a relative sense, i.e., it permits to identify changes qualitatively, but cannot be used to obtain exact quantitative data. The calculated model values – of around 270 nm for the ethene-exposed crystals – agree with the dimensions obtained by evaluating the corresponding SEM images of both, as-synthesized and treated samples, shown in Fig. 3.2. As it can be observed, the crystals still show clear defined cubic structures with no observable change in morphology when compared to the untreated sample. For the ethane experiment, the model suggested a coherent scattering domain of around 220 nm, which is below what is observed in the SEM images. Nonetheless, when comparing SEM images – prior and after treatment – the size of the crystals remains consistently around 400 nm. Furthermore, the shapes of the crystals are still clearly defined and no edge degradation is observable after ethane exposure. (Note: Observable cracks in the images were produced by the SEM beam when measuring the samples, while uncovered areas are attributed to the breaking of the support which could compromise the surface and therefore interfere in the achievement of a defect-free film; the calculated Williamson and Hall plots can be found in Appendix A.4).

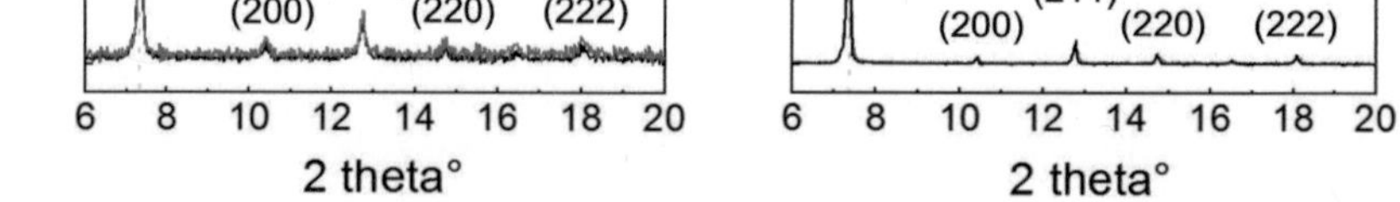

Figure 3.1: Comparison of normalized XRD patterns of ZIF-8 SURMOF crystals exposed to a constant stream of a) ethene or b) ethane for 24 h at room temperature

Finally, the normalized IRRA spectra of as-synthesized and treated membranes at room temperature were also semi-quantitatively analyzed; the modes at 1384 cm^{-1} and 1584 cm^{-1} corresponding to the methyl bending and C=N stretching assignments, respectively, were compared. As shown in Tab. 3.1, the room temperature scenarios' ratios of the intensities of both modes are close to 1, further supporting that no compositional changes happened to the crystals neither for exposure to ethene nor ethane. The slight deviations over 1 are attributed to possible artifacts during the baseline correction step.

Based on all the obtained data, it could be concluded that the established LPE-LBL ZIF-8 SURMOF system is stable at room temperature for both gases.

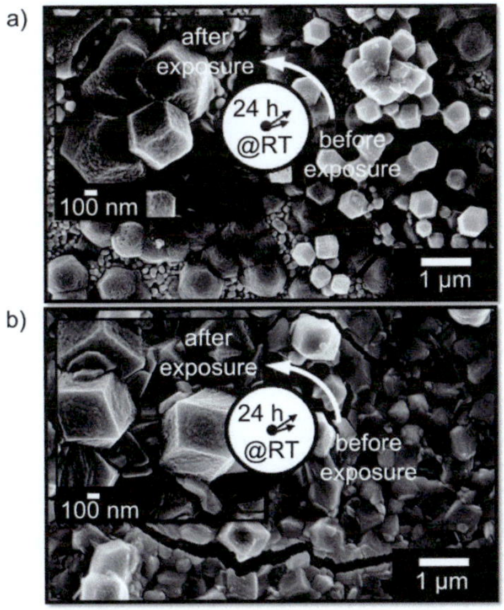

Figure 3.2: SEM images of ZIF-8 SURMOF crystals after a) ethene or b) ethane exposure for 24 h at room temperature with a background image showing the crystals before treatment

a) Ethene

	Ratios			
Mode [cm^{-1}]	$\frac{\text{treated}_{@ RT}}{\text{as-syn.}}$	$\frac{\text{treated}_{@ 100°C}}{\text{as-syn.}}$	$\frac{\text{treated}_{@ 150°C}}{\text{as-syn.}}$	$\frac{\text{treated}_{@ 200°C}}{\text{as-syn.}}$
1384	1.16	0.79	0.80	0.50
1584	1.06	0.87	0.76	0.83

b) Ethane

	Ratios			
Mode [cm^{-1}]	$\frac{\text{treated}_{@ RT}}{\text{as-syn.}}$	$\frac{\text{treated}_{@ 100°C}}{\text{as-syn.}}$	$\frac{\text{treated}_{@ 150°C}}{\text{as-syn.}}$	$\frac{\text{treated}_{@ 200°C}}{\text{as-syn.}}$
1384	1.02	0.96	1.08	0.01
1584	1.12	1.13	0.95	0.18

Table 3.1: Intensity ratios of the modes at 1384 cm^{-1} and 1584 cm^{-1} for the normalized IRRA spectra between as-synthesized and a) ethene- or b) ethane-exposed ZIF-8 SURMOF crystals at diverse temperatures for 24 h

3.2.2.2. Higher temperature scenarios

Seeking to further determine the temperature dependency of the established LPE-LBL ZIF-8 SURMOF system, more stability treatment experiments were conducted. First of all, the treated samples were measured with XRD to observe any change in crystallinity. As shown in Fig. 3.3 and as discussed in the last section, no change is detectable between as-synthesized and room temperature treated crystals for either gas. As the temperature starts to increase, slight changes are observed in the diffractograms, specially for the ones exposed to ethene. Though the crystallinity remains in both cases up to 150 °C, it is clearly observable that the peak intensity increases for all planes relatively to that of the (110) plane, i.e., the orientation of the crystals in the (110) plane is reduced. This last point is of high importance for gas separation applications, since a predominant orientation on the (110) is highly desired to provide facilitated access to guest molecules. For the 200 °C scenarios, the diffractrograms showed an increase in noise and no defined peaks, for which its normalized pattern could not be included in Fig. 3.3; both corresponding XRD patterns can be found in Appendix A.4.

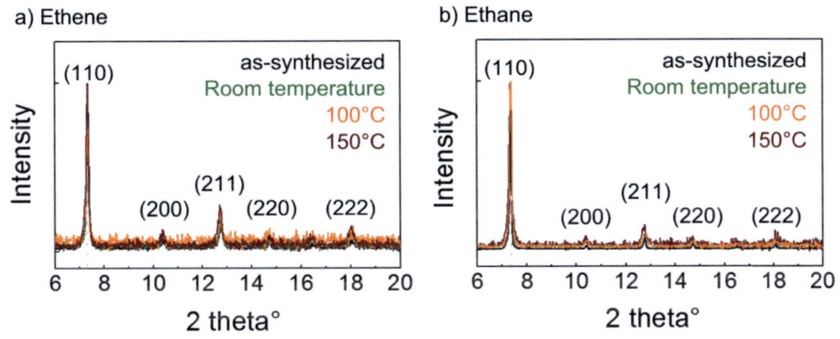

Figure 3.3: Comparison of normalized XRD patterns of ZIF-8 SURMOF crystals after a) ethene or b) ethane exposure for 24 h at room temperature, 100 °C, and 150 °C

Following, pre- and after treatment IRRA spectra were compared as shown in Fig. 3.4 for ethene and Fig. 3.5 for ethane.

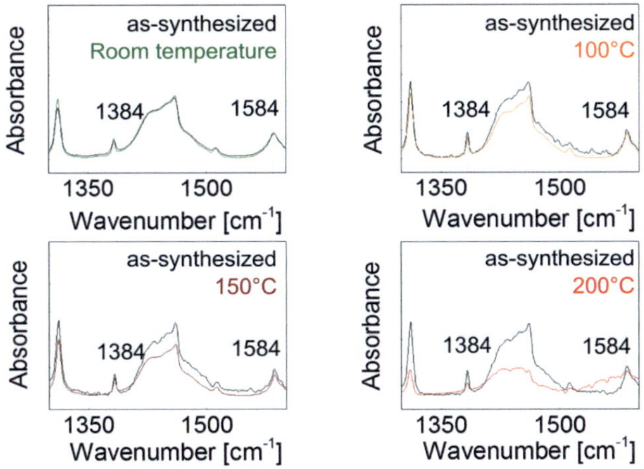

Figure 3.4: Normalized IRRA spectra of ZIF-8 SURMOF crystals after ethene exposure for 24 h at room temperature, 100 °C, 150 °C, and 200 °C

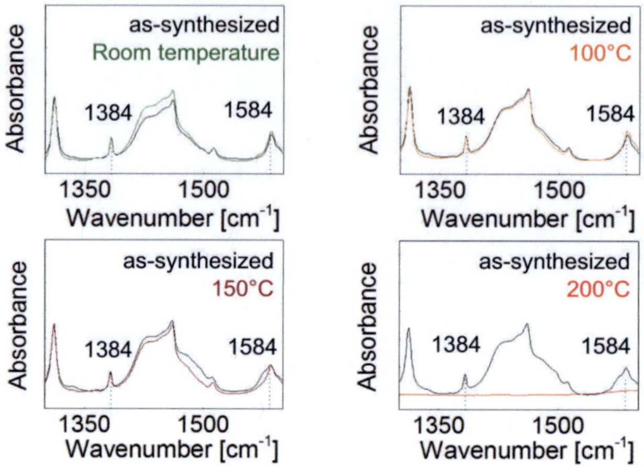

Figure 3.5: Normalized IRRA spectra of ZIF-8 SURMOF crystals after ethane exposure for 24 h at room temperature, 100 °C, 150 °C, and 200 °C

Furthermore, the ratios between treated and as-synthesized normalized IR bands at 1384 cm^{-1} and 1584 cm^{-1} corresponding to the methyl bending and C=N stretching assignments respectively, were compared and reported in Tab. 3.1. As mentioned previously, for the room temperature scenarios the values of the ratios are near 1, indicating no significant change in chemical composition under either ethene or ethane exposure. Continuing with higher temperature scenarios, the ethene-exposed crystals showed a clear decreasing trend in the values of the ratios when going above 100 °C, which is in agreement with the observed broadening of the bands probably due to C=N bond cleavage followed by a demethylation. For the ethane-exposed crystals, the ratios remained close to 1 even up to 150 °C, suggesting higher stability of the ZIF-8 crystals under this gas. Nonetheless when reaching the 200 °C, the composition was undoubtedly severely affected for both cases, agreeing with the XRD results obtained. Important and intriguing to observe is that though the crystals showed higher stability under ethane up to 150 °C, when reaching 200 °C the temperature effect was more intense and no IR modes were detectable, clearly reflected on the ratio values. Note: The slight deviations over 1 are attributed to possible artifacts during the baseline correction step.

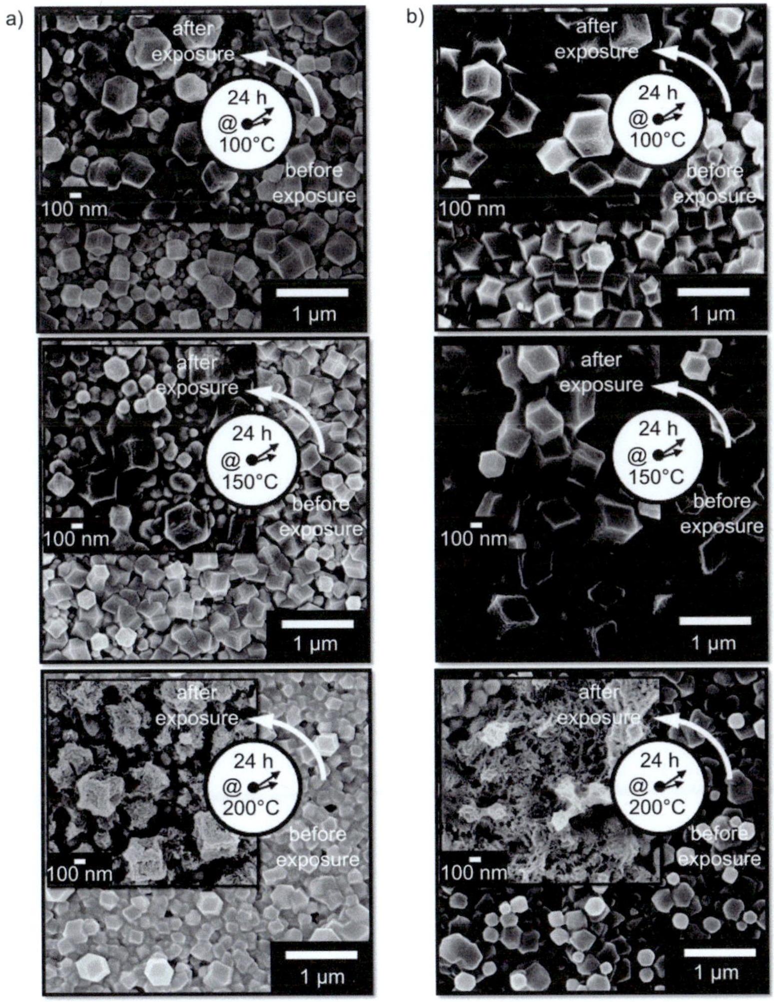

Figure 3.6: SEM images of ZIF-8 SURMOF crystals after a) ethene or b) ethane exposure for 24 h at 100 °C, 150 °C, and 200 °C with a background image showing the crystals before treatment

Finally to support the obtained data, SEM images were obtained for the as-synthesized and treated ZIF-8 SURMOF samples at 100 °C, 150 °C, and 200 °C as shown in Fig. 3.6. The sample exposed at 100 °C under ethene, showed no significant changes in crystal size ($\sim$ 200 nm) compared to the as-synthesized sample. Also the cubic structure could still be clearly identified, though some small effects on the surface of them began to be visible. For ethane, no effect was observable, leaving the cubic crystals intact and of the same size as the as-synthesized sample. As the temperature continued rising, effects could be observed in both cases. Cubic structures started decomposing and merging with one another which indicated the beginning of the formation of amorphous structures. For the 200 °C scenarios, the cubic structures were no longer completely distinguishable, showing great degradation in edges and surfaces. Furthermore, it appears that up to 150 °C, the ZIF-8 SURMOF crystals are more stable under ethane than under ethene, but when going above 150 °C the degradation effect is more pronounced for the ethane-exposed crystals; more data is still required to corroborate such behavior.

Based on all the previous data obtained, it can be concluded that the established LPE LBL ZIF-8 SURMOF system remained stable under ethene and ethane exposure up to 100 °C. However, as the temperatures started rising above this point, changes in crystallinity, composition, and morphology could be detected as it was confirmed with XRD, IRRAS, and SEM. For the samples tested at 200 °C, both cases resulted in an X-ray amorphous phase and a significant decrease in the IR intensity of the characteristic ZIF-8 modes.

3.3. Compression Tests

Likewise, determining the stability of the established LPE LBL ZIF-8 SURMOF system under different pressures is of importance. It was relevant to this study to make sure that the system remains stable at least below P_{abs} =1.5 bar, corresponding to the maximum operation range used during the characterization and gas separation experiments. Nonetheless, just as in the past section, being able to determine the limit of the system is also of high interest to identify further possible applications.

3.3.1. Experimental Methodology

Figure 3.7: Compression setup using a static testing device Z.05 including a load cell with a nominal load force of 200 N from Zwick

In order to study the pressure effect on the LPE LBL ZIF-8 crystals, several membranes were synthesized with 175 deposition cycles following the procedure previously described in Chapter 2. Ideal testing would involve the mounting of the membrane inside the gas separation module and constant flow of gas at diverse pressures. Nonetheless, an insertion and removal of the glued membrane, as well as absolute pressures above 1.5 bar are not possible in the actual gas separation modules and setup, respectively. Therefore, using a static testing device Z.05 with a load cell of 200 N nominal force, type XforceP 2mV/V from Zwick, ZIF-8 SURMOF membranes (cross-sectional area of $1.327 \cdot 10^{-4}$ m^2) were mounted and compressed by applying forces of 6.5 N ($P_{abs} \approx 1.5$ bar), 50 N ($P_{abs} \approx 4.78$ bar), and 100 N ($P_{abs} \approx 8.5$ bar) over the membranes' surface as shown in Fig. 3.7. The continuous compression forces were applied for 60 s followed by 5 s relaxation during a 1.5 h period seeking to observe any effect on the crystallinity of the material given higher pressure conditions over a defined operation time. It is important to mention that this test would permit the evaluation of any immediate change in crystallinity, but could not be used to identify the pressure limit of the system with respect to crack formations, since the force applied with this device is not the same type of force applied by the gas during the gas separation tests. XRD measurements were performed before and after the compression tests, followed by SEM investigation.

3.3.2. Results and discussion

In Fig. 3.8 the XRD patterns for the compressed samples are reported. As it can be observed, the crystallinity prevailed in all cases, showing even a slightly higher orientation towards the (110) plane. This can be related to the known ZIF-8 stimuli-driven structural rearrangements probably caused by the rotation of the methyl-1H-imidazole units in the framework which can occur for instance via pressure changes or by the presence of guest molecules [13, 104]. This point has become highly attractive for research, as it can explain the unexpected entrance of larger molecules through the theoretically smaller ZIF-8 pore apertures, expanding the application possibilities of this material [105, 106].

To support the XRD data, SEM images were obtained for each of the previous scenarios as shown in Fig. 3.9. As it can be observed, in all of the scenarios the cubic structures are still completely observable agreeing with the reported crystallinity of the XRD patterns. No crystal appears to be broken, scratched, or significantly affected in its morphology. As explained previously, the creation of crack and/or defects is of high interest, but unfortunately may not be proven via this method, since the pressure applied differs (i.e., not even all over the membrane unless having a perfectly homogeneous surface) from the one applied during the gas separation experiments. It is still interesting to mention that for the 6.5 N and

50 N scenarios, some slight cracks are visible probably due to an uneven surface seeking rearrangement when being compressed. Surprisingly though, the 100 N scenarios showed no cracks and highly defined crystal structures.

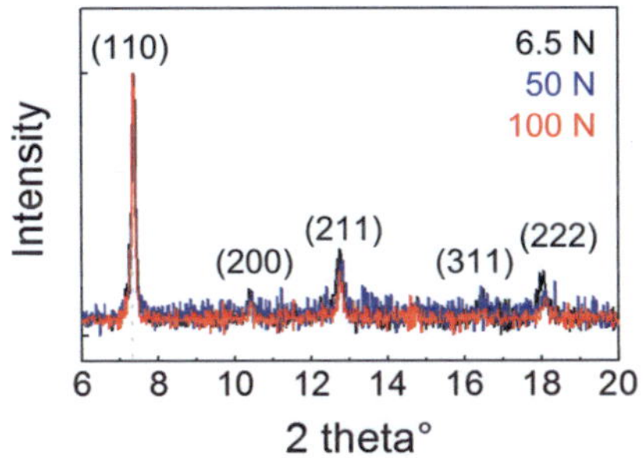

Figure 3.8: Comparison of normalized XRD patterns of ZIF-8 SURMOF membranes compressed by forces of a) 6.5 N, b) 50 N, and c) 100 N during 1.5 h

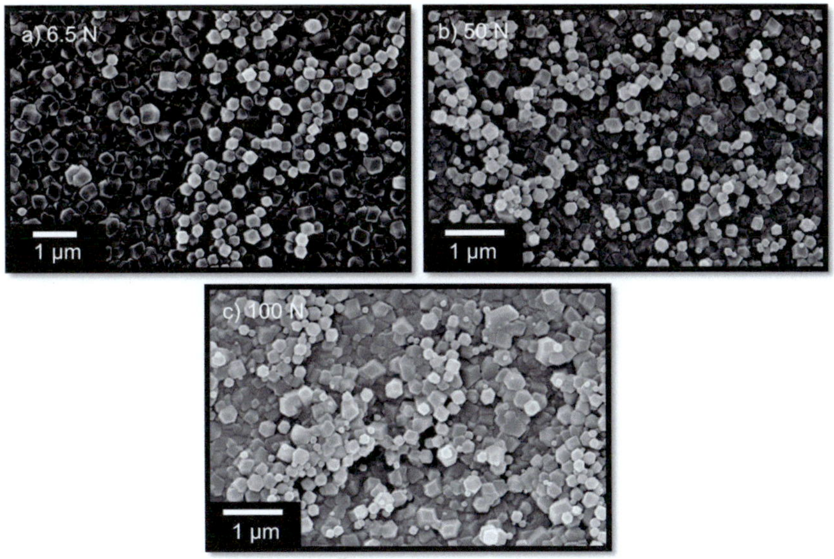

Figure 3.9: SEM images of ZIF-8 SURMOF membranes compressed by forces of 6.5 N, 50 N, and 100 N during 1.5 h

Based on the data, it could be confirmed that the crystallinity of the LPE LBL ZIF-8 SURMOF crystals remained even when being compressed under a 100 N force. Nonetheless, the formation of cracks and/or defects could not be determined via this method. Taking the performed permeation experiments at P_{abs} =1.5 bar or below into consideration, the membranes showed the same behavior even within measurement spans of several months and measurement times involving 10 continuous hours. This suggests a long term stability of the membranes, at least within the pressure regime used for the experiments mentioned above.

4. ZIF-8 SURMOF membrane-based alkane/alkene gas separation

Membrane technologies are an attractive less-energetically demanding alternative for a variety of separation processes. When speaking about their application in gas separation, diverse manufacturing challenges are known. This last point though, has also driven great research in the field expecting to achieve highly stable, permeable, and selective membranes [107].

Due to some exclusive features MOFs possess, including tunability and immobilization of functional sites, the optimization of their sieving effects as well as their ability to differentiate their interactions with gas molecules have been demonstrated, respectively [108].

As described in Chapter 1, ZIFs are a popular MOF subclass characterized for its porous, flexible, and thermally stable structures [107]. Within this subclass, diverse topologies and pore apertures have been reported, opening the gates to various promising applications [109]; up-to-date this subclass corresponds to the most studied within the gas separation field [17]. Within the subclass, ZIF-8 excels as the most promising candidate. This ZIF compound is highly attractive due to its large inner cage and small window apertures ranging within the kinetic diameters of several gas molecules [38].

In this chapter, the gas separation performance of the synthesized LPE LBL ZIF-8 SURMOF membranes deposited on Au-coated α-Al_2O_3 supports was studied for various binary feed gas mixtures composed of ethene and ethane. Currently, the selectivity of ZIF-8 in C_2H_4/C_2H_6 systems is known to be limited, reaching only separation factors (SFs) between 2, as reported by Shekhah et al. [13] and James et al. [89], and 2.8 as reported by Bux et al. [91] depending on the feed pressure conditions. The goal of this chapter was to determine the system's performance in comparison to other known synthesis techniques; different feed compositions and pressures were used and all experiments were conducted using the Wicke-Kallenbach method. The results provided general knowledge of the actual separation capacities which could lead to further target-wise designed systems. Single feed gas permeation experiments were also performed. The obtained experimental data was then used to estimate the M-S surface diffusivities of ethene and ethane required in the simulation study presented in Chapter 5.

4.1. Binary feed gas mixture separation

4.1.1. Experimental methodology

The gas separation experiments in this work were performed using the Wicke-Kallenbach technique firstly introduced in 1941 [110] and commonly used for its characteristic advantage concerning the elimination of forced flow [111]. Samples deposited on Au-coated α-Al_2O_3 supports for 175 cycles, which were previously presented in Chapter 2 as samples 1-4, were studied; the same color code is presented, i.e., sample 1 is represented in orange, sample 2 in green, sample 3 in purple, and sample 4 in burgundy. These synthesized membranes were found to possess an overall promising quality as presented in Chapter 2 and were therefore studied for gas separation. In Tab. 4.1, the binary feed experiments performed with the aforementioned samples are presented.

Scenario	Feed Composition	Sweep gas
Equimolar	$30 \frac{ml}{min}$ C_2H_4 + $30 \frac{ml}{min}$ C_2H_6	$60 \frac{ml}{min}$ He
Ethene-rich (2:1)	$40 \frac{ml}{min}$ C_2H_4 + $20 \frac{ml}{min}$ C_2H_6	$60 \frac{ml}{min}$ He
Ethane-rich (1:2)	$20 \frac{ml}{min}$ C_2H_4 + $40 \frac{ml}{min}$ C_2H_6	$60 \frac{ml}{min}$ He

Table 4.1: Binary feed gas separation experiments performed in this work

An illustration of the set-up is presented in Fig. 4.1. The feed composed of ethene and ethane controlled by MFCs, entered the module through the upper chamber, while helium was introduced as sweep gas through the lower chamber of the module in order to flush the permeate side imposing a concentration gradient [112]. Assuring negligible absolute pressure difference over the membrane, the experiments were conducted at absolute pressures of 1.1, 1.2, and 1.3 bar (equal on both chambers [113]). The membrane was left in constant operation at the assigned pressure until it reached a steady state, which was reflected with constant flow and concentration measurements. A flow meter was used to determine the total flow on the retentate side, while the permeate outlet composition was analyzed via a GC Chromatograph GC_ 7890B from Agilent Technologies equipped with HP-Molesieve and HP-Plot Q columns from Agilent J&W GC Columns. The stabilization time differed from experiment to experiment, but was in average reached within 2-3 hours of continuous operation.

One of the greatest disadvantages of the Wicke-Kallenbach method corresponds to the counter-diffusion of the sweep gas towards the retentate side, which complicates the situation and could affect the permeation of the feed side components [3]. Nevertheless, this method is commonly employed due to its suitability for transient measurements [114]. Important to consider though, is that the measurements based on such experiments, e.g., diffusion coefficients, might be lower than the true pure-component values [115]. Therefore, for the sake of consistency and considering the effect such counter-diffusion could have in the values obtained, both single and binary feed gas measurements in this work were performed using the same technique.

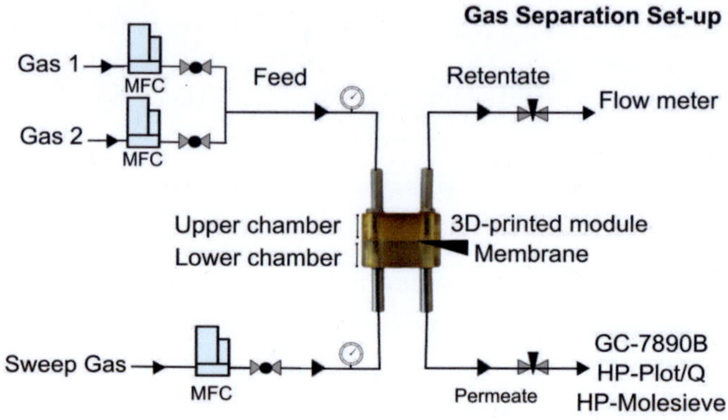

Figure 4.1: Gas separation set-up sketch based on the Wicke-Kallenbach principle using a 3D-printed module composed of an upper and a lower chamber containing the ZIF-8 SURMOF membrane (adapted from [35])

4.1.2. Performance parameters: C_2H_4/C_2H_6 separation in ZIF-8 systems

The performance of each of the studied samples was evaluated via the separation factor (SF) defined as:

$$\alpha_{i,j} = \frac{y_i}{y_j} \cdot \frac{x_j}{x_i} \tag{4.1}$$

where y and x correspond to the molar fractions of component i and j in the permeate and retentate sides, respectively [116]. In this work ethene is referred to as component i and ethane as component j.

The permeances for ethene and ethane were also obtained. As mentioned in the experimental methodology, different feed compositions and pressures were investigated. According to various literature, the mixture of ethene and ethane favors ethane adsorption within the ZIF-8 pores. Therefore, a known preferential adsorption of ethane competes with a preferential diffusion of ethene in ZIF-8 [89, 91, 117], which complicates the separation performance of the membranes. According to these same literature, the preferential diffusion of ethene is said to overcompensate the preferential adsorption of ethane. Therefore, the study of variation in feed composition was of interest to this work in order to define whether the membrane might perform differently under certain operation conditions.

The effect of different pressures was also investigated. On the published work by Bux et al. [91], slightly different C_2H_4/C_2H_6 separation factors were reached depending on the feed pressure, going from SFs of 2.8 to 2.4 when operating at 1 and 6 bar feed pressure, respectively. Theoretically, such slight variations in the SFs might be associated to the induced rotational displacement of the imidazole ligands with pressure variations [89]. In this study, only low absolute pressure scenarios (1.1-1.3 bar) were tested and no significant effect could be observed within such pressure range. Therefore, the following analysis is based primarily on the variation of the feed composition at an absolute pressure of 1.1 bar. The permeance graphs covering the different pressures can be found in Appendix A.5.

4.1.3. Results and discussion

The results of the performed experiments are shown in Fig. 4.2 and Tab. 4.2. As it can be observed, all of the membranes exhibited ethene selectivity, but the effect given the feed composition differed from membrane to membrane. In the following paragraphs, the individual results of each of the measured samples is discussed. Afterwards, a comparison between all of the samples can be found, seeking to further understand the role diverse factors play in the ethene and ethane transport through the ZIF-8 pores.

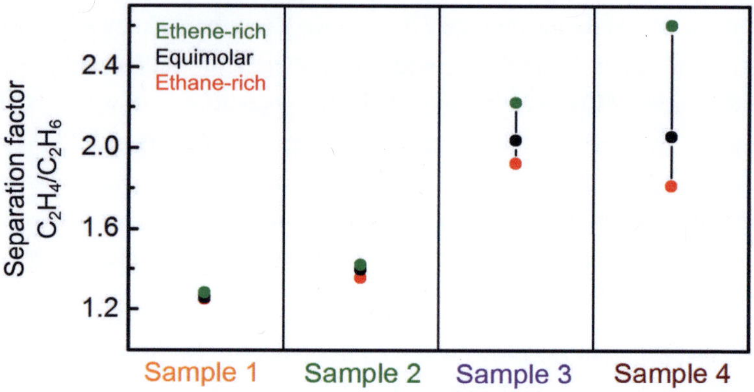

Figure 4.2: Separation factors of samples 1-4 (ZIF-8 SURMOF membranes) for different feed compositions at P_{abs} =1.1 bar

Scenario	Gas	Sample 1	Sample 2	Sample 3	Sample 4
		Permeance [10^{-7}mol•m^{-2}•s^{-1}•Pa^{-1}]			
Equimolar	Ethene	6.99	2.85	1.59	0.43
	Ethane	5.21	1.98	0.76	0.21
Ethene-rich	Ethene	6.86	2.90	1.58	0.40
	Ethane	4.98	1.97	0.69	0.15
Ethane-rich	Ethene	6.92	2.77	1.52	0.37
	Ethane	5.19	1.98	0.77	0.20

Table 4.2: Ethene and ethane permeance values of samples 1-4 (ZIF-8 SURMOF membranes) for different feed compositions at P_{abs} =1.1 bar

Sample 1 For this membrane, the highest C_2H_4/C_2H_6 SF was achieved in the ethene-rich scenario being just under 1.30, quite below the expected value when compared to other literature sources. Nonetheless, as it might be recalled from Chapter 2 Fig. 2.13, sample 1 showed a slight upward profile and the highest H_2 permeance of all four samples. This suggested sample 1 as the membrane possessing the greatest number of defects and/or the largest ones. After measurements, sample 1 showed indeed the worst performance of all samples. Such defects greatly affected the selectivity of the membrane favoring to a certain degree non-selective transport, though still exhibiting a slight selectivity towards ethene. Furthermore given the defective conditions of the membrane, the different tested scenarios and pressures played no role, exhibiting the membrane always C_2H_4/C_2H_6 SFs between 1.25 and 1.28, with ethene permeances between 6.86 and $6.99 \cdot 10^{-7} mol \cdot m^{-2} \cdot s^{-1} \cdot Pa^{-1}$, and ethane permeances between 4.98 and $5.21 \cdot 10^{-7} mol \cdot m^{-2} \cdot s^{-1} \cdot Pa^{-1}$.

Sample 2 This membrane exhibited a similar behavior to sample 1, i.e., the different feed compositions and pressures played no major role in the membrane's performance. A slightly higher ethene selectivity was again observable in the ethene-rich scenario, reaching a separation factor of 1.46. As it might be recalled from Chapter 2, sample 2 had the second highest H_2 permeance of all four samples, though the difference in comparison to sample 1 was quite considerable (45-55 %). This is believed to be associated with the presence of less micro-defects, which given the same experimental conditions could be related to synthesis artifacts regarding concentrations, humidity conditions, support variations, and/or support handling. Overall, the performance of the membrane improved, but was still below reported C_2H_4/C_2H_6 SFs of ZIF-8 membranes. The SFs ranged between 1.35 and 1.46 depending on the scenario, with ethene permeances between 2.77 and $2.90 \cdot 10^{-7} mol \cdot m^{-2} \cdot s^{-1} \cdot Pa^{-1}$, and ethane permeances between 1.97 and $1.98 \cdot 10^{-7} mol \cdot m^{-2} \cdot s^{-1} \cdot Pa^{-1}$.

Sample 3 For sample 3, the performance of the membrane significantly improved and the scenarios started to play an important role in the behavior of the membrane, as it can be observed in Fig. 4.2. Under equimolar feed composition, the membrane exhibited a C_2H_4/C_2H_6 SF of 2.04 comparable to diverse literature [13, 89, 100]. When tested under an ethene-rich feed, the SF improved reaching a C_2H_4/C_2H_6 SF of 2.22, while lowering for the ethane-rich scenario down to 1.92. The ethene and ethane permeances were also below the ones for samples 1 and 2 ranging between 1.52 and $1.59 \cdot 10^{-7}$ mol·m^{-2}·s^{-1}·Pa^{-1} for ethene, and between 0.69 and $0.77 \cdot 10^{-7}$ mol·m^{-2}·s^{-1}·Pa^{-1} for ethane.

Sample 4 The measurements of sample 4 delivered the same C_2H_4/C_2H_6 SF as sample 3 under equimolar feed, regardless its decrease in ethene and ethane permeance. Nonetheless, the effect on the SFs with varying feed composition was more significant than with sample 3. The membrane exhibited ethene selectivity reaching a SF of 2.60 with an ethene-rich feed, while lowering its performance down to 1.81 in the ethane-rich scenario. The ethene and ethane permeances were the lowest of all samples ranging between 0.37 and $0.43 \cdot 10^{-7}$ mol·m^{-2}·s^{-1}·Pa^{-1} and between 0.15 and $0.21 \cdot 10^{-7}$ mol·m^{-2}·s^{-1}·Pa^{-1}, respectively.

Comparison of samples As already mentioned, all of the samples exhibited ethene selectivity in the gas separation experiments, but differed in the separation factor and in the ethene and ethane permeance values. Furthermore, an interesting point was also observed among the membranes regarding the feed composition variations. As clearly depicted in Fig. 4.2, samples 1 and 2 showed low C_2H_4/C_2H_6 SFs, suggesting the presence of defects within the synthesized surface. This was also supported by the higher H_2 permeances measured and reported in Chapter 2 in comparison to samples 3 and 4. Though samples 1 and 2 exhibited low ethene selectivities, their defects favored non-selective transport resulting in similar results regardless of the feed composition. On the other hand, samples 3 and 4 performed differently depending on the feed scenario. Nevertheless, the effect was more compelling in sample 4. This is thought to be related to the lower permeance of the membrane possibly possessing less grain boundary micro-defects [11, 13] thus favoring the enhancement or drop in the C_2H_4/C_2H_6 SFs. Furthermore, this can also be related to the known trade-off in membranes between permeability and selectivity. Given similar thicknesses in samples 3 and 4, the higher selectivities of sample 4 would agree with the lower permeability of the membrane. Fig. 4.2 demonstrates how micro-defects may still greatly influence a membrane's performance.

Counter-flux As mentioned at the beginning of the chapter, the biggest disadvantage when using the Wicke-Kallenbach method is the possible counter-diffusion of the sweep gas to the retentate side, which would lower the permeation of the feed components. Such effect is often considered negligible, but such assumption has been proven to depend on different factors, including the nature of the sweep gas, the feed composition, and the temperature of operation [115]. Considering these points, helium was selected as the sweep gas for the experiments conducted in this work, given its negligible adsorption in the pressure range used in this work as reported in literature [104], as well as its molecular weight and size when compared to other possible sweep gases, e.g., argon. The counter-flux of helium was nevertheless quantified and analyzed for each of the studied samples.

In Fig. 4.3 the relationship between the ethane partial pressure difference and the counter-flux $\%$ of the sweep gas, i.e., the percentage of helium that permeated to the retentate chamber from the permeate chamber, is depicted for the different feed compositions. In the literature, it has been repeatedly reported that ZIF-8 exhibits an ethane preferential adsorption selectivity over ethene [91, 117]. Relating this fact to the counter-diffusion of helium in the Wicke-Kallenbach experiments, it is expected that molecules that adsorb more strongly help reduce such effect [115]. As the ZIF-8 pores would be occupied by the adsorbing component(s), the sweep gas counter-diffusion would be expected to be mainly through the non-ZIF-8 pores [118], i.e., defects. This point is supported in Fig. 4.3. As the partial pressure difference of ethane lowered in the different scenarios, the counter-flux of helium increased and viceversa. Also, for a He counter-flux above 25$\%$, i.e., samples 1 and 2, no compelling variation could be observed as the feed composition was varied. As the counter-flux lowered, i.e., samples 3 and 4, different behaviors could be then recognized. Furthermore, the decrease of the helium counter-flux with increasing ethane partial pressure difference can also be explained by the increased friction exerted by the ethane molecules on the molecules inside the ZIF-8 pores. Important to mention is that the experimental configuration was the same for all of the measured samples, meaning the counter-flux of helium was not manipulated in any way and instead was defined by the membranes themselves.

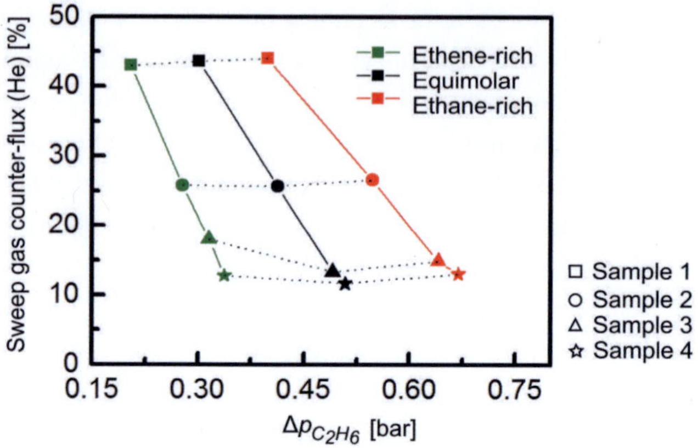

Figure 4.3: Sweep gas (He) counter-flux dependence on ethane partial pressure difference for various feed composition scenarios

In general, this graph supported once again that samples 1 and 2 possess larger micro-defects, and that the poor performance of these membranes might be also due to the helium back permeation, proper of the Wicke-Kallenbach method. When analyzing further these variations, the ethene-rich feed generally presented the highest counter-flux percentage, followed then by the ethane-rich and equimolar composition respectively, upholding once again ethane as the stronger-adsorbing component hindering the counter-diffusion of helium in the studied system. This point can be further related to the findings by Do and Do [119], which reported this so-called cooperation/competition effect in systems similar to ZIF-8. This and similar works generally claim that, in what have been called cation-free nonpolar pore systems, a competition between a stronger-adsorbing component (ethane) and a weaker-adsorbing component (ethene) can be observed. The behavior of the adsorption isotherms of the gases in the mixture state therefore differ from the single component ones, specifically speaking, paraffin adsorption is favored over olefin adsorption depending on the feed composition [91, 119, 120].

To further analyze this effect, the ethane partial pressure on the retentate chamber was plotted against the ethane permeation flux as reported in Fig. 4.4. First of all, it is clear how the permeation flux increased as the membrane overall quality lowered, i.e., the presumed amount of micro-defects present on the membrane. Also, as the ethane partial pressure increased, so did the ethane permeation flux (given the increased partial pressure difference). The interesting point to observe though is the behavior with composition. In samples 1, 2, and 3 going from an ethane-rich feed to an equimolar one, or from an equimolar feed to an ethene-rich one resulted in similar changes. Sample 4 on the other hand, experienced a lower change when going from an ethane-rich feed to an equimolar one, which might suggest some difference in behavior with feed composition on the sample with lower amount of presumed micro-defects. In order to reach proper conclusions on this point, more membranes and experimental results are required, including the determination of defects on each membrane.

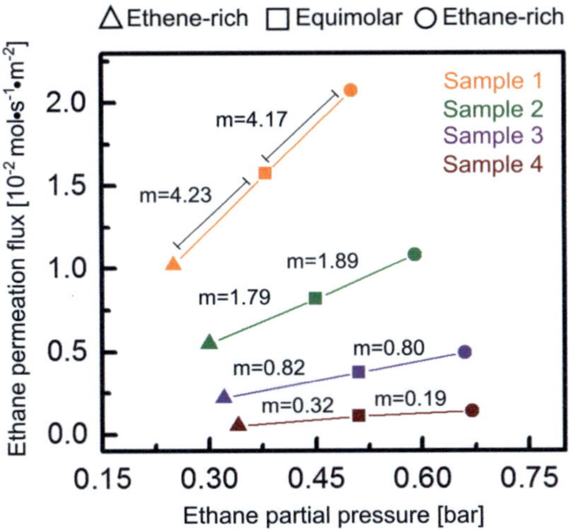

Figure 4.4: Ethane partial pressure vs. ethane permeation flux for various feed composition scenarios of samples 1-4 (ZIF-8 SURMOF membranes) with "m" representing the slope between the two points

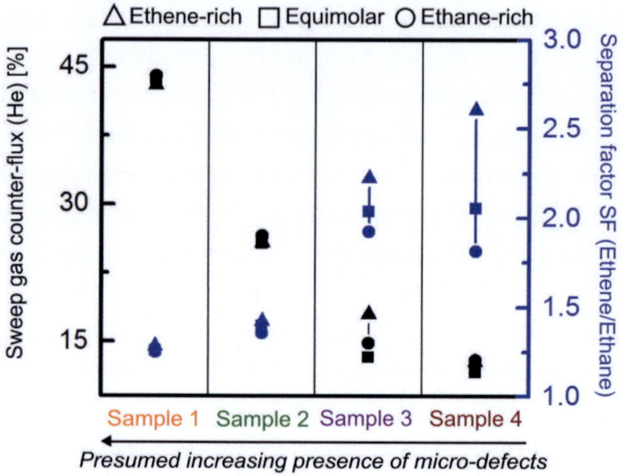

Figure 4.5: He counter-flux and separation factors of samples 1-4 (ZIF-8 SURMOF membranes) for different feed compositions at P_{abs} =1.1 bar

As a closing part to this section, the counter-flux percentage of helium (black axis) and the separation factor (blue axis) are compared among the studied membranes for different feed compositions as depicted in Fig. 4.5. As it can be observed, as the presumed presence or size of micro-defects lowered, so did the counter-flux of the sweep gas. Furthermore, as mentioned in previous sections a significant improvement in performance could be observed for samples 3 and 4. Relating the permeance values to this graph, samples 1 and 2 reported permeances between 2.77 and $6.99 \cdot 10^{-7} mol \cdot m^{-2} \cdot s^{-1} \cdot Pa^{-1}$ for ethene and between 1.97 and $5.21 \cdot 10^{-7} mol \cdot m^{-2} \cdot s^{-1} \cdot Pa^{-1}$ for ethane. Samples 3 and 4 reported permeances between 0.37 and $1.59 \cdot 10^{-7} mol \cdot m^{-2} \cdot s^{-1} \cdot Pa^{-1}$ for ethene and between 0.15 and $0.77 \cdot 10^{-7} mol \cdot m^{-2} \cdot s^{-1} \cdot Pa^{-1}$ for ethane.

4.2. Single feed gas permeation experiments

Considering the previous results in the binary feed gas section, single feed gas permeation experiments were conducted on samples 3 and 4 in order to obtain experimental data required to calculate the M-S surface diffusivities for each of the studied gases. Obtaining these parameters was of great importance for the simulation study to be discussed in Chapter 5.

4.2.1. Experimental methodology

Using the same set-up and procedure presented in the binary feed gas mixture section, single feed gas permeation experiments were performed on samples 3 and 4. In this case, the feed was composed only of ethene or ethane; helium was once again used as sweep gas. Tab. 4.3 shows the single feed gas permeation experiments conducted at P_{abs} =1.1 bar.

Single gas	Feed Composition	Sweep gas
Ethene	$30\frac{ml}{min}$ C_2H_4	$30\ \frac{ml}{min}$ He
Ethane	$30\frac{ml}{min}$ C_2H_6	$30\ \frac{ml}{min}$ He

Table 4.3: Single feed gas permeation experiments performed in this work

4.2.2. Results and discussion

4.2.2.1. M-S surface diffusivities calculation

Parameter definition As described in Chapter 1 Eq. (1.20), the M-S surface diffusivity is defined as:

$$D_i^s = -\frac{J_i^s}{\rho_m \cdot q_{sat} \cdot \nabla\theta_i} \cdot (1 - \overline{\theta_i}).$$

For the system studied in this work, ρ_m is equal to 924,250 g$\cdot$ m^{-3} [44]. J_i^s, $\nabla\theta_i$, and $\overline{\theta_i}$ were calculated for each gas based on the information obtained from the single feed gas permeation experiments.

	Ethene		Ethane		
	k	q_{sat}	k	q_{sat}	Pressure range
	[bar^{-1}]	[mmol•g^{-1}]	[bar^{-1}]	[mmol•g^{-1}]	[bar]
Pub. 1	0.3622	6.1934	0.1668	7.4124	0-8
Pub. 2	0.1674	9.9906	0.2155	14.3969	0-1
Pub. 3	0.468	7.0125	1.2276	6.1143	0-8
Pub. 4	0.1	11.8	0.26	8.25	0-4

Table 4.4: k and q_{sat} values fitted from extracted literature: publication 1 [13], publication 2 [86], publication 3 [121], and publication 4 [104]

Two parameters are still required in order to calculate the M-S diffusivities: q_{sat} and k. Corresponding adsorption isotherms found in the works by 1) Shekhah et al. [13], 2) Böhme et al. [86], and 3) Wu et al. [121] were fitted to the Langmuir model using the nonlinear least-squares solver from MATLAB® in order to obtain q_{sat} and k; the work by 4) Zhang et al. [104] already provided the required parameters. The fitted graphs are reported in Fig. 4.6. This fitting procedure is of great importance. If the functional form does not fit the data, non-representative results may be delivered resulting in inaccurate predictions [122, 123]. Furthermore, the consulted adsorption data was also compared. For ethene adsorption slight differences could be observed, but for ethane the variations within the reported data were more pronounced. For this reason, it was preferred to use each literature source independently. A summary of the obtained fitted data and reported parameters is shown in Tab. 4.4.

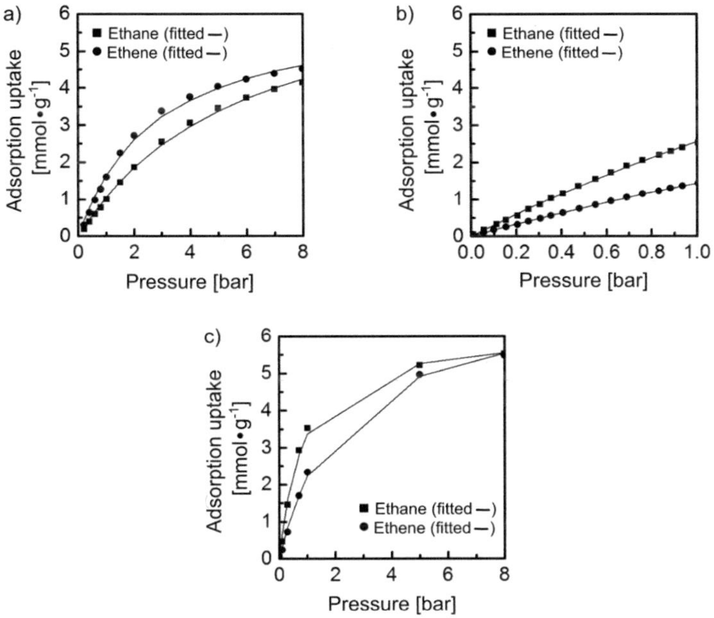

Figure 4.6: Adsorption data (symbols) reported in the works by a) Shekhah et al. [13], b) Böhme et al. [86], and c) Wu et al. [121] fitted (solid line) to the Langmuir model using the nonlinear least-squares solver from MATLAB®

Results Single feed gas permeation experiments were conducted using samples 3 and 4. For sample 3, the experiment resulted in a flux (J^s) of 11.7 mmol·m^{-2}·s^{-1}, while delivering a J^s of 5.8 mmol·m^{-2}·s^{-1} for ethane. Sample 4, which exhibited lower permeances in the gas separation experiments, consistently delivered lower permeating fluxes and permeate partial pressures. The fluxes recorded were of J^s of 3.9 mmol·m^{-2}·s^{-1} for ethene and of J^s of 2.4 mmol·m^{-2}·s^{-1} for ethane.

Following, θ_r and θ_p were calculated using Eq. (1.11). $\overline{\theta_i}$ corresponds to the average of both θ_r and θ_p, while $\nabla\theta$ corresponds to the gradient of the coverage of ethene or ethane along the membrane thickness. Based on SEM images, a membrane thickness of 500 nm for 175 deposition cycles experiments was assumed in all further calculations. The calculated parameters for ethene and ethane of samples 3 and 4 are shown in Tab. 4.5.

a)

	Sample 3							
	Ethene				Ethane			
	Θ_R	Θ_P	$\bar{\Theta}$	$\nabla\Theta$ [m^{-1}]	Θ_R	Θ_P	$\bar{\Theta}$	$\nabla\Theta$ [m^{-1}]
Pub. 1	0.2544	0.0332	0.1438	-442429.1	0.1324	0.0086	0.0705	-247504.0
Pub. 2	0.1362	0.0156	0.0759	-241194.3	0.1646	0.0111	0.0879	-307097.8
Pub. 3	0.3060	0.0425	0.1742	-527015.8	0.5289	0.0601	0.2945	-937705.6
Pub. 4	0.0861	0.0094	0.0477	-153407.1	0.1921	0.0134	0.1027	-357511.3

b)

	Sample 4							
	Ethene				Ethane			
	Θ_R	Θ_P	$\bar{\Theta}$	$\nabla\Theta$ [m^{-1}]	Θ_R	Θ_P	$\bar{\Theta}$	$\nabla\Theta$ [m^{-1}]
Pub. 1	0.2670	0.0112	0.1391	-511643.4	0.1418	0.0032	0.0725	-277098.2
Pub. 2	0.1441	0.0052	0.0747	-277801.2	0.1759	0.0042	0.0900	-343448.1
Pub. 3	0.3201	0.0144	0.1673	-611258.6	0.5487	0.0233	0.2860	-1050859.1
Pub. 4	0.0914	0.0031	0.0473	-176543.5	0.2048	0.0050	0.1049	-399502.9

Table 4.5: Calculated θ_r, θ_p, $\overline{\theta_i}$, and $\nabla\theta_i$ for a) sample 3 and b) sample 4 using single feed gas experimental data from this work in combination with the k and q_{sat} values from diverse literature [13, 86, 104, 121]

Finally, the M-S surface diffusivity was calculated for samples 3 and 4 using the obtained experimental data, as well as the diverse literature sources. Though within similar ranges, the diffusivities in each of the cases differed. When considering the ratios within the diffusivities, the results using publications 2, 3, and 4 were within reasonable ranges, while publication 1 obtained almost the same diffusivity for both gases. Possible reasons for these variations are the different synthesis methods used that could play an important role in the material's properties, for instance the presence or absence of certain micro-defects. For the literature data reported by Shekhah et al. [13] and for the work by Böhme et al. [86], ZIF-8 solvothermal synthesis as reported in the work by Park et al. [19] was performed. The data reported by Wu et al. [121] consisted in GCMC simulated pure-component isotherms, and the parameters reported in the work by Zhang et al. [104] were obtained using commercial BASF ZIF-8. All of the obtained single gas M-S diffusivities were tested in the simulation study to be discussed next in Chapter 5. A comparison of these obtained data is also presented in more detail further on.

	Sample 3			Sample 4		
	D^s $[m^2 \cdot s^{-1}]$		$D^s_{Ethene}/D^s_{Ethane}$	D^s $[m^2 \cdot s^{-1}]$		$D^s_{Ethene}/D^s_{Ethane}$
	Ethene	Ethane		Ethene	Ethane	
Publication 1	$3.96 \cdot 10^{-12}$	$3.21 \cdot 10^{-12}$	1.2	$1.15 \cdot 10^{-12}$	$1.15 \cdot 10^{-12}$	1.0
Publication 2	$4.86 \cdot 10^{-12}$	$1.31 \cdot 10^{-12}$	3.7	$1.42 \cdot 10^{-12}$	$0.47 \cdot 10^{-12}$	3.0
Publication 3	$2.83 \cdot 10^{-12}$	$0.78 \cdot 10^{-12}$	3.6	$0.82 \cdot 10^{-12}$	$0.28 \cdot 10^{-12}$	2.9
Publication 4	$6.67 \cdot 10^{-12}$	$1.93 \cdot 10^{-12}$	3.5	$1.94 \cdot 10^{-12}$	$0.69 \cdot 10^{-12}$	2.8

Table 4.6: Calculated M-S surface diffusivities using single feed gas experimental data from this work in combination with the k and q_{sat} values from diverse literature [13, 86, 104, 121]

4.2.2.2. Ideal selectivity (permselectivity)

After performing the single feed gas permeation experiments, the ideal selectivities or permselectivities of the studied membranes were calculated. This parameter corresponds to the ratio of the single gas permeabilities of two different gas components under (otherwise) identical conditions [124]:

$$S_{ij} = \frac{\mathfrak{p}_i}{\mathfrak{p}_j} \tag{4.2}$$

where S_{ij} is the ideal selectivity of component i to j, and $\mathfrak{p}_i$ and $\mathfrak{p}_j$ correspond to the single gas permeabilities of component i and j, respectively. This parameter is helpful in providing a hint to the performance of the membrane in a binary mixture scenario. Nonetheless, once in a mixture the molecule interactions between the various components can play an important role and affect the transport within the membrane, influencing the selectivity. For this reason, the separation factor already presented at the beginning of this chapter, is instead used to describe the real separation ability of a membrane for mixed components [124, 125].

In Tab. 4.7 the ideal selectivities obtained from the single feed gas permeation experiments are reported for samples 3 and 4; the separation factors presented in the previous section are also included for the different feed compositions. As it can be observed, sample 3 reported basically the same selectivity in the equimolar composition scenario, while sample 4 actually reported a better separation than expected when compared to its ideal selectivity. Samples 3 and 4 were synthesized using the same procedure and basically under the same conditions, but still exhibited a difference in the permeance values measured, which greatly influenced their performance. Furthermore, the counter-flux effect on both, single and binary feed gas experiments, is something that must be also kept in mind. This point is analyzed and discussed in greater detail in the following chapter.

	Permselectivity $S_{C_2H_4/C_2H_6}$	Separation factor ($SF_{C_2H_4/C_2H_6}$)		
		Equimolar	Ethene-rich	Ethane-rich
Sample 3	2.04	2.04	2.22	1.92
Sample 4	1.67	2.05	2.60	1.81

Table 4.7: Calculated permselectivities from single feed gas permeation experiments and separation factors from binary feed gas separation experiments for different feed compositions of samples 3 and 4

4.3. Reflection and open questions

In conclusion to this chapter, samples 3 and 4 resulted in high quality LPE LBL ZIF-8 SURMOF membranes with possible micro-defects, achieving good separation results in agreement with the reported performance of other ZIF-8 systems. Though the principle has been proven, the SFs are still below the minimum requirements in order to consider any possible industrial application.

Exploiting now the known tunability capacity of MOFs could help design target-wise systems which should outperform the actual LPE LBL ZIF-8 SURMOF membranes in ethene/ethane separation and/or other gas mixtures.

However, the performance results must be further analyzed in order to obtain further information and answer several intriguing questions from the studied system. First of all, a general study of the transport mechanism through the ZIF-8 pores is sought. Second, after analyzing the behavior of several parameters with varying feed composition, a reduction in the sweep gas counter-flux was observed with a higher presence of ethane. This reduction was also dependent on the permeance of the membrane which can be related to the presence of grain boundary micro-defects. So, how do these variations affect the adsorption interactions and therefore the transport in ZIF-8 pores and with it its separation performance?

Furthermore, the counter-flux of the sweep gas during the single and binary feed gas permeation experiments must be further discussed. As mentioned previously, the Wicke-Kallenbach technique possesses this disadvantage which can be somehow minimized depending on certain factors including the nature of the sweep gas. Considering the effect such counter-diffusion could have in the values obtained, both single and binary feed gas measurements in this work were performed with the same method seeking to obtain consistent parameters. Still, this can only be considered as an approximation and does not determine the role/effect the counter-flux played in the separation capability of the studied membranes and thus in the data collected.

Pursuing an answer to such questions and reflections, a simulation was implemented and presented in the next chapter. Understanding the experimental data through theory might provide a rich insight into different parameters playing key roles in the ZIF-8 SURMOF system's transport, which could eventually aid in the design of a target-wise system.

5. Simulation model adapted for a ZIF-8 SURMOF system

As explained in Chapter 1, diverse mass transport mechanisms in the gas phase can be identified within porous media. The research on this field has a long history [70], dating back to the beginning of the 20th century with the works of Knudsen [126, 127] and Smoluchowski [128, 129]. Still, many questions remain for every system has its own properties and therefore its own transport behavior under concentration and pressure gradients, or even adsorption fields. Nevertheless, several models have been proposed and proven to work within certain systems sharing common features.

When speaking about multicomponent mass transfer in porous media, one of the most known proposed models corresponds to the generalized Maxwell-Stefan equations of Krishna and co-workers [76], which have been successfully applied in diverse works [76, 78]. This model has been proven to deliver in many cases better results when compared for instance to Fick's law, due to the proper consideration of thermodynamic non-idealities, as well as the influence of external force fields [72]. As explained in Chapter 1, "the basis of the Maxwell-Stefan theory is that the driving force for movement that is acting on a species is balanced by the friction that is experienced by that species [78]." The model can treat transport in adsorbed state, i.e., considering the adsorption equilibrium of (gas) mixtures. For the latter, an adsorption model must be selected in order to describe the surface coverage and thus the chemical potential gradient.

The goal of this chapter was to be able to link the experimental data collected and presented in the previous chapters with theory seeking to understand the transport mechanism through the ZIF-8 pores, as well as to discuss several open questions regarding the adsorption interactions and permeance values influence on the membrane's separation performance. Using MATLAB®, the generalized M-S surface diffusion model was tested via two different adsorption models: IAST and extended Langmuir. The approach described next was developed in the works by Ding [59, 130] and adapted to the specific needs of this system.

5.1. Simulation scheme and workflows

5.1.1. Approach description

The general approach of the simulation consists in the discretization of the ZIF-8 film in different layers, n_l. As depicted in Fig. 5.1, each of the layers has two interfaces, each with its own molar fractions ($y_1(i)$; $y_2(i)$ $\forall$ $i \in$ $[1, n+1]$), surface coverages ($\theta_1(i)$; $\theta_2(i)$ $\forall$ $i \in$ $[1, n+1]$), and adsorbed fractions ($x_1(i)$; $x_2(i)$ $\forall$ $i \in$ $[1, n+1]$); this last one only when using IAST. The values in these interfaces are then used to obtain the average values of the molar fractions ($y_{ave1}(j)$; $y_{ave2}(j)$ $\forall$ jß $[1, n]$) and the surface coverages ($\theta_{ave1}(j)$; $\theta_{ave2}(j)$ $\forall$ $j \in$ $[1, n]$) of each layer considering linear behavior within each of the interfaces. Based on these, the molar fluxes for each component can be then simulated and compared to the experimental data collected. For the sake of consistency within the simulation runs and text in this chapter, subscript 1 will always refer to ethene, subscript 2 to ethane, and subscript 3 to the sweep gas used, i.e., helium. For the ZIF-8 SURMOF film, the generalized M-S surface diffusion model from Krishna et al. [76] was used; the support's resistance was also considered in the simulation. A detailed explanation concerning the criteria and assumptions for each of these can be found in the following section.

5.1.1.1. ZIF-8 SURMOF layer

Model and assumptions To model this layer, the Maxwell-Stefan surface diffusion model was implemented. As explained in Chapter 1, this model has been proven to accurately describe the multicomponent mass transfer in porous systems [131]. The model analyzes the diffusion of binary gas mixtures at the nanopore level considering the friction between moving molecules and the adsorption sites (molecule-wall interactions) characterized by D_i^s, i.e., M-S surface diffusivity of component i, as well as the friction between adsorbed molecules moving through the pores (molecule-molecule interactions) characterized by D_{ij}^s, i.e., the cross-term surface diffusivity between component i and j in adsorbed state. For more detailed information on the theory behind this model, please refer to Chapter 1.

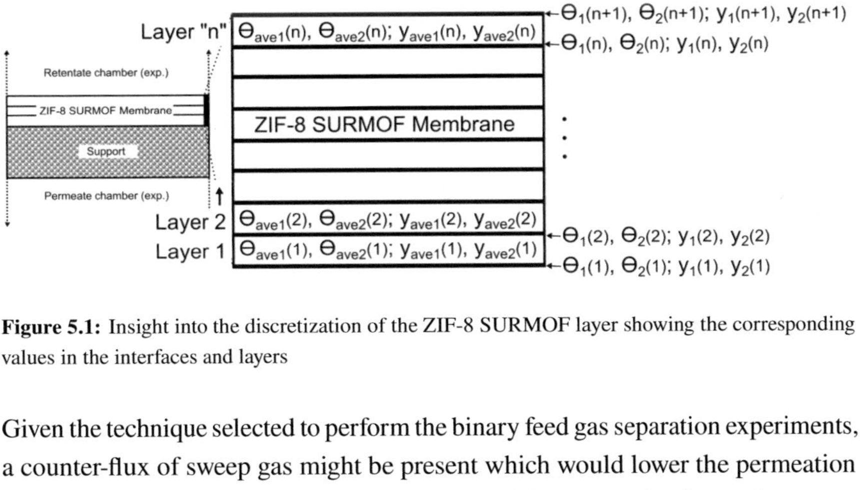

Figure 5.1: Insight into the discretization of the ZIF-8 SURMOF layer showing the corresponding values in the interfaces and layers

Given the technique selected to perform the binary feed gas separation experiments, a counter-flux of sweep gas might be present which would lower the permeation of the studied components. Depending on several factors, neglecting such counter-flux could play a significant role in the simulation. However in order to incorporate this phenomena extensive experimental work is required. Also as previously mentioned, the counter-flux effect has been proven to depend on different factors, including the nature of the sweep gas, the feed composition, and the temperature of operation [115]. According to literature [104], a negligible adsorption of helium in the ZIF-8 framework at the pressure study in this work (P_{abs} =1.1 bar) can be then assumed.

Furthermore, the pressure-dependent experiments presented in Chapter 2 supported that regardless of using the same experimental conditions, the membranes exhibited different H_2 permeance values, which can be associated to the presence of micro-defects in the ZIF-8 SURMOF layer. This implies that within the layer not only ZIF-8 pores are present, but also non-ZIF-8 pores, i.e., micro-defects. Considering ZIF-8 narrow size windows of 3.4 Å and the H_2 permeance values in Fig. 2.13, it can be assumed that the diffusion through these larger pores will be somewhat different, though eventually qualitatively similar [78]. Additionally, the exact size of such defects is unknown to this work, though efforts are currently being done for a possible incorporation of such in future works.

Therefore, for simplicity reasons and considering the above points, the counter-flux of the sweep gas as well as the contribution of the defects in the transport were not considered in the hereby presented simulation.

Finally, as shown in Eq. (1.9) and Eq. (1.10), the driving force for diffusion is the chemical potential gradient which is linked to the gradient in surface coverage by a matrix of thermodynamic factors for which the selection of an adsorption model is required [76]. In this work, two different adsorption models were investigated: ideal adsorbed solution theory (IAST) and extended Langmuir. Their implementation in the simulation are described in detail in the following sections. To learn more about their theoretical background, please refer to Chapter 1.

5.1.1.2. Au-coated α-Al_2O_3 support layer

Model and assumptions The consideration of the support layer, i.e., its resistance, was also implemented in the simulation, permitting a better understanding of the behavior at the interface between the ZIF-8 SURMOF and the Au-coated α-Al_2O_3 support.

Though in the Wicke-Kallenbach method no absolute pressure difference is presented, i.e., no viscous flow phenomena [114], and small volumetric flows were employed for the experiments in this work, a partial-pressure gradient, which should not be neglected, exists across the Au-coated α-Al_2O_3 support. As introduced in the works by van de Graaf [78, 115] given the similar support's properties (please refer to Chapter 2) and experimental conditions [132], the support's resistance as a result of molecular diffusion can be described as:

$$-\frac{1}{RT}\nabla p_i = \sum_{j=1}^{n} \frac{y_j J_i - y_i J_j}{\varepsilon_s \cdot D_{ij}^{BF}} \qquad i, j = 1, ..., n, \tag{5.1}$$

where ε_s is the support's porosity and D_{ij}^{BF} are the binary friction diffusivities, which can be calculated using the empirical correlation developed by Fuller, Schettler, and Giddings [66]:

$$D_{ij}^{BF} = \frac{10^{-3} \cdot T^{1.75}(\frac{1}{M_i} + \frac{1}{M_j})^{0.5}}{P((\sum_i \upsilon_i)^{1/3} + (\sum_j \upsilon_i)^{1/3})^2} \tag{5.2}$$

where D_{ij}^{BF} is in $cm^2{\cdot}s^{-1}$, T is in K, M is in $g{\cdot}mol^{-1}$, P is in atm, and υ refers to the atomic diffusion volumes; these experimental diffusion data is tabulated in the work by Fuller, Schettler, and Giddings [66].

The corresponding diffusivities were calculated considering a temperature of 296.15 K and a pressure of $1.1{\cdot}10^5$ Pa with 1) representing ethene, 2) ethane, and 3) helium:

- $D_{1,2}^{BF} = 9.980 \cdot 10^{-6}$ [$m^2{\cdot}s^{-1}$]
- $D_{1,3}^{BF} = 4.199 \cdot 10^{-5}$ [$m^2{\cdot}s^{-1}$]
- $D_{2,3}^{BF} = 4.002 \cdot 10^{-5}$ [$m^2{\cdot}s^{-1}$].

Important to mention is that three components were included in the calculations: ethene, ethane, and the sweep gas. Nevertheless similar to the ZIF-8 layer, the counter-flux of helium through the support was considered negligible [78]. To explain this assumption, it might be recalled from previous chapters that both, single and binary feed gas permeation experiments, were conducted using the Wicke-Kallenbach method. Considering the experimental data obtained and that both experiments were performed using the same technique, the incorporation of this effect was only indirectly considered via the experimental data rather than explicitly in the simulation, i.e. only as a first approximation. A more thorough analysis of this assumption is presented later on in the chapter. Finally a linear profile was assumed between the end of the support and its interface with the ZIF-8 layer.

5.1.2. Simulation workflows

A general overview of the program's workflow for each of the two adsorption models tested is shown in Fig. 5.2. As it can be seen, the first step of the program consists in calling input parameters which are listed in Tab. 5.1. The next step consists in defining the adsorption model, i.e., IAST or extended Langmuir. Following different steps are undergone depending on the model chosen; a detailed explanation for each are presented next.

5.1.2.1. IAST

As shown in Fig. 5.2, the algorithm's goal is to solve for $y_1(i)$; $y_2(i)\,\forall\, i \in [1, n{+}1]$. This is achieved by using the Levenberg-Marquardt algorithm implemented in the *fsolve* function from the MATLAB® Optimization Toolbox®. To start with the process, the necessary adsorbed fractions are also determined using the Levenberg-Marquardt algorithm, as well as the molar fraction initial guesses.

With the obtained adsorbed fractions, the coverage values are then estimated with the IAST model (please refer to Chapter 1). Using the coverage and molar fraction values at the interfaces, the average coverage and molar fractions in each layer are then calculated; linear profiles are assumed. Following, the cross-term surface diffusivities are estimated using the Vignes empirical relation. The support's resistance is then solved. The optimization algorithm stops when the mass balances reach the established tolerance. Using the simulated molar fraction values, the molar fluxes ($J_1^{s(sim)}$, $J_2^{s(sim)}$) can now be estimated. The obtained simulated molar fluxes can be now compared to the experimental data.

5.1.2.2. Extended Langmuir

The procedure for this adsorption model is a bit more simplified since adsorbed fractions are not required. Once again, the goal is to solve for $y_1(i);\ y_2(i)\ \forall\ i \in [1,\ n+1]$ using the Levenberg-Marquardt algorithm implemented in the *fsolve* function from the MATLAB® Optimization Toolbox®. First of all, the coverage values are estimated using the extended Langmuir equation (please refer to Chapter 1). From then on, the exact same procedure as for IAST (cf. 5.1.2.1) follows.

Input Parameter	Symbol	Value
Number of components	comp.	2
Number of discretized layers	n_l	1, 2, 3, 4, 8
Pressure of system	P	$1.15 \cdot 10^5$ [Pa]
Temperature of system	T	296.15 [K]
Molar masses of components	M	$M_1 = 28.05 \cdot 10^{-3}$; $M_2 = 30.07 \cdot 10^{-3}$ [$kg \cdot mol^{-1}$]
Universal gas constant	R	8.314 [$J \cdot mol^{-1} \cdot K^{-1}$]
Density of ZIF-8 SURMOF framework	ρ_m	924.25 [$kg \cdot m^{-3}$]
Thickness of membrane	δ_m	$5 \cdot 10^{-7}$ m
Thickness of support	δ_s	$1 \cdot 10^{-3}$ m
Porosity of support	ε_s	0.5
Binary friction diffusivities	D^{BF}	$D^{BF}_{1,2} = 9.980 \cdot 10^{-6}$; $D^{BF}_{1,3} = 4.199 \cdot 10^{-5}$; $D^{BF}_{2,3} = 4.002 \cdot 10^{-5}$ [$m^2 \cdot s^{-1}$]
M-S surface diffusivities	D^S	Refer to Table 4.6
Adsorption constants	k	Refer to Table 4.4
Saturation capacities	q_{sat}	
Molar fractions in retentate	y_r	Experimental data (corresponding scenario and membrane)
Molar fractions in permeate	y_p	Experimental data (corresponding scenario and membrane)

Table 5.1: Program's input parameters, symbols, and corresponding values where 1 is ethene, 2 is ethane, and 3 is helium (sweep gas)

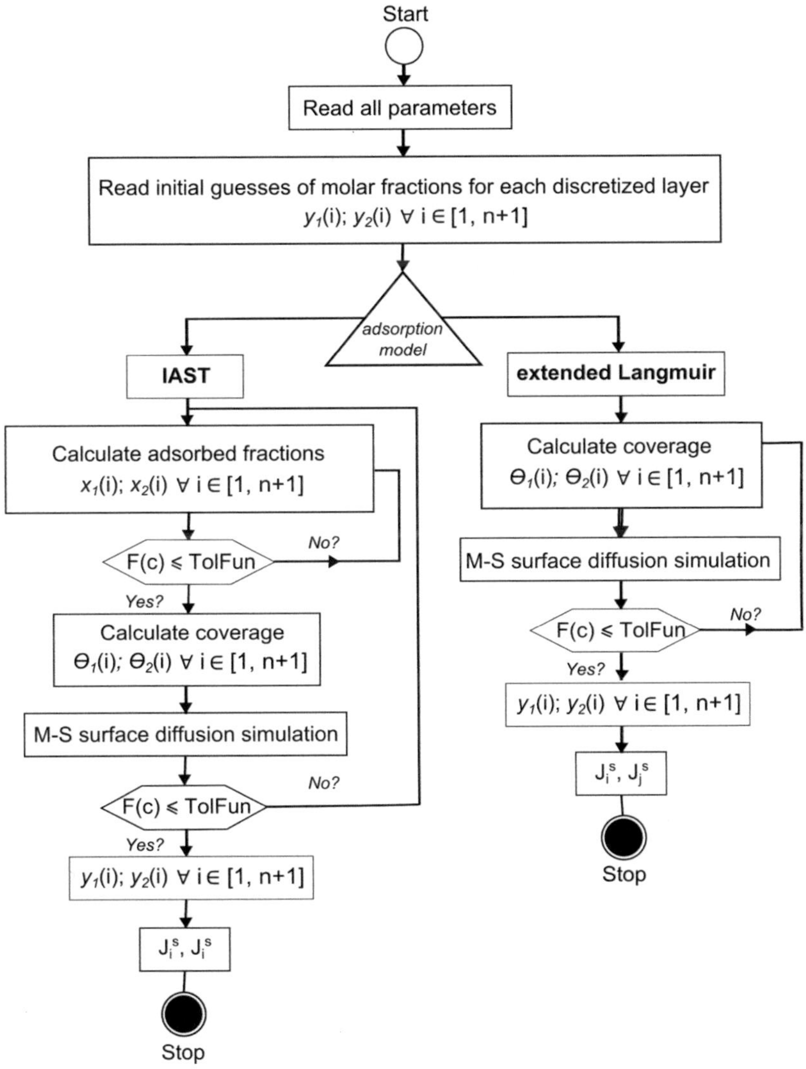

Figure 5.2: Program's workflow for simulation of the ZIF-8 SURMOF system including two adsorption models: IAST and extended Langmuir

5.2. Results and discussion

Similar to the gas separation experiments presented in Chapter 4, three different feed composition scenarios were also considered in the simulation runs: equimolar, ethene-rich, and ethane-rich. Each scenario was tested a total of 8 times: adsorption data consulted from 4 different literature sources (see Tab. 4.4) and 2 different adsorption models. This procedure was conducted twice, first for sample 3 and then for sample 4. A preliminary value of n_l=8 was used for all of the runs seeking to obtain an accurate profile; this parameter was revised and presented later on in this chapter. The results obtained are reported in Tab. 5.2 and Tab. 5.3 corresponding to samples 3 and 4, respectively. The tables show the simulated molar fluxes of each component in each of the different scenarios tested compared to the real experimental values. The percent variations between the simulated and experimental data are also reported.

At first glance, the simulation results seem to favor to a great extent the implementation based on the extended Langmuir adsorption model. The considerable percent variations obtained for the IAST runs were unexpected, considering the results published in the work by Bux et al. [91], where the IAST model was able to describe accurately the mixture adsorption of ethene and ethane in ZIF-8 when compared to the GCMC-IRM mixture isotherms in different feed composition scenarios. However, it could be observed that though the fluxes were underestimated when using the IAST adsorption model, the ratios of the fluxes were in agreement with the experimental ones.

		Equimolar			Ethene-rich			Ethane-rich		
		$J_1^{s(sim)}$	$J_2^{s(sim)}$	$\frac{J_1^{s(sim)}}{J_2^{s(sim)}}$	$J_1^{s(sim)}$	$J_2^{s(sim)}$	$\frac{J_1^{s(sim)}}{J_2^{s(sim)}}$	$J_1^{s(sim)}$	$J_2^{s(sim)}$	$\frac{J_1^{s(sim)}}{J_2^{s(sim)}}$
IAST	Pub. 1	4.03	1.91	2.11	5.32	1.28	4.16	2.70	2.56	1.05
	Pub. 2	3.62	2.04	1.77	4.92	1.38	3.57	2.35	2.70	0.87
	Pub. 3	3.05	2.59	1.18	4.52	1.82	2.48	1.82	3.28	0.55
	Pub. 4	3.49	2.07	1.69	4.81	1.42	3.39	2.23	2.71	0.82
extended Langmuir	Pub. 1	7.49	3.58	2.09	9.85	2.38	4.14	5.05	4.83	1.05
	Pub. 2	6.69	3.95	1.69	9.25	2.67	3.46	4.24	5.18	0.82
	Pub. 3	4.58	4.53	1.01	7.09	3.29	2.16	2.66	5.60	0.48
	Pub. 4	6.21	4.00	1.55	8.78	2.76	3.18	3.90	5.19	0.75
		$J_1^{s(exp)}$	$J_2^{s(exp)}$	$\frac{J_2^{s(exp)}}{J_1^{s(exp)}}$	$J_1^{s(exp)}$	$J_2^{s(exp)}$	$\frac{J_2^{s(exp)}}{J_1^{s(exp)}}$	$J_1^{s(exp)}$	$J_2^{s(exp)}$	$\frac{J_2^{s(exp)}}{J_1^{s(exp)}}$
Experimental		7.44	3.73	2.00	9.40	2.17	4.33	4.64	4.93	0.94
		$\frac{J_1^{s(sim)}}{J_1^{s(exp)}}-1$	$\frac{J_2^{s(sim)}}{J_2^{s(exp)}}-1$	$\frac{J_1^{s(sim)}/J_2^{s(sim)}}{J_1^{s(exp)}/J_2^{s(exp)}}-1$	$\frac{J_1^{s(sim)}}{J_1^{s(exp)}}-1$	$\frac{J_2^{s(sim)}}{J_2^{s(exp)}}-1$	$\frac{J_1^{s(sim)}/J_2^{s(sim)}}{J_1^{s(exp)}/J_2^{s(exp)}}-1$	$\frac{J_1^{s(sim)}}{J_1^{s(exp)}}-1$	$\frac{J_2^{s(sim)}}{J_2^{s(exp)}}-1$	$\frac{J_1^{s(sim)}/J_2^{s(sim)}}{J_1^{s(exp)}/J_2^{s(exp)}}-1$
IAST	Pub. 1	-46%	-49%	6%	-43%	-41%	-4%	-42%	-48%	12%
	Pub. 2	-51%	-45%	-11%	-48%	-36%	-18%	-49%	-45%	-8%
	Pub. 3	-59%	-30%	-41%	-52%	-16%	-43%	-61%	-33%	-41%
	Pub. 4	-53%	-44%	-16%	-49%	-35%	-22%	-52%	-45%	-13%
extended Langmuir	Pub. 1	1%	-4%	5%	5%	10%	-4%	9%	-2%	11%
	Pub. 2	-10%	6%	-15%	-2%	23%	-20%	-9%	5%	-13%
	Pub. 3	-38%	22%	-49%	-25%	51%	-50%	-43%	14%	-50%
	Pub. 4	-17%	7%	-22%	-7%	27%	-27%	-16%	5%	-20%

Table 5.2: Sample 3: simulated and experimental molar fluxes, J, [mmol·m^2·s^{-1}] and ratios with their corresponding percent variation for different feed scenarios where 1 is ethene and 2 is ethane

		Equimolar			Ethene-rich			Ethane-rich		
		$J_1^{s(sim)}$	$J_2^{s(sim)}$	$\frac{J_1^{s(sim)}}{J_2^{s(sim)}}$	$J_1^{s(sim)}$	$J_2^{s(sim)}$	$\frac{J_1^{s(sim)}}{J_2^{s(sim)}}$	$J_1^{s(sim)}$	$J_2^{s(sim)}$	$\frac{J_1^{s(sim)}}{J_2^{s(sim)}}$
IAST	Pub. 1	1.24	0.69	1.80	1.63	0.46	3.54	0.84	0.93	0.90
	Pub. 2	1.12	0.74	1.51	1.52	0.50	3.04	0.73	0.98	0.74
	Pub. 3	0.97	0.95	1.02	1.41	0.67	2.10	0.59	1.21	0.49
	Pub. 4	1.08	0.75	1.44	1.48	0.51	2.90	0.70	0.99	0.71
extended Langmuir	Pub. 1	2.33	1.29	1.81	3.05	0.85	3.59	1.58	1.75	0.90
	Pub. 2	2.10	1.43	1.47	2.87	0.97	2.96	1.35	1.89	0.71
	Pub. 3	1.52	1.67	0.91	2.29	1.20	1.91	0.90	2.08	0.43
	Pub. 4	1.96	1.45	1.35	2.74	1.00	2.74	1.25	1.89	0.66
		$J_1^{s(exp)}$	$J_2^{s(exp)}$	$\frac{J_1^{s(exp)}}{J_2^{s(exp)}}$	$J_1^{s(exp)}$	$J_2^{s(exp)}$	$\frac{J_1^{s(exp)}}{J_2^{s(exp)}}$	$J_1^{s(exp)}$	$J_2^{s(exp)}$	$\frac{J_1^{s(exp)}}{J_2^{s(exp)}}$
Experimental		2.18	1.06	2.05	2.67	0.52	5.15	1.24	1.37	0.90
		$\frac{J_1^{s(sim)}}{J_1^{s(exp)}}-1$	$\frac{J_2^{s(sim)}}{J_2^{s(exp)}}-1$	$\frac{J_1^{s(sim)}/J_2^{s(sim)}}{J_1^{s(exp)}/J_2^{s(exp)}}-1$	$\frac{J_1^{s(sim)}}{J_1^{s(exp)}}-1$	$\frac{J_2^{s(sim)}}{J_2^{s(exp)}}-1$	$\frac{J_1^{s(sim)}/J_2^{s(sim)}}{J_1^{s(exp)}/J_2^{s(exp)}}-1$	$\frac{J_1^{s(sim)}}{J_1^{s(exp)}}-1$	$\frac{J_2^{s(sim)}}{J_2^{s(exp)}}-1$	$\frac{J_1^{s(sim)}/J_2^{s(sim)}}{J_1^{s(exp)}/J_2^{s(exp)}}-1$
IAST	Pub. 1	-43%	-35%	-12%	-39%	-11%	-31%	-32%	-32%	0%
	Pub. 2	-49%	-30%	-26%	-43%	-4%	-41%	-41%	-28%	-17%
	Pub. 3	-55%	-11%	-50%	-47%	29%	-59%	-52%	-12%	-46%
	Pub. 4	-50%	-29%	-30%	-45%	-2%	-44%	-43%	-28%	-22%
extended Langmuir	Pub. 1	7%	21%	-12%	14%	64%	-30%	28%	28%	0%
	Pub. 2	-4%	34%	-28%	7%	87%	-43%	9%	38%	-21%
	Pub. 3	-30%	57%	-56%	-14%	131%	-63%	-27%	52%	-52%
	Pub. 4	-10%	36%	-34%	3%	93%	-47%	1%	38%	-27%

Table 5.3: Sample 4: simulated and experimental molar fluxes, J, [mmol·m^2·s^{-1}] and ratios with their corresponding percent variation for different feed scenarios where 1 is ethene and 2 is ethane

In order to reach proper conclusions, different points were considered. First of all, the different adsorption literature sources were compared. Second, the known existing counter-flux of the sweep gas – proper of the Wicke-Kallenbach technique – and its effect on the experimental measurements was also further revised. Furthermore, the varying feed composition effect on the simulation results was also studied. Finally, given the different permeance values (i.e., presumed amount of defects) measured for samples 3 and 4, a comparison of the simulation results of both membranes is of interest to qualitatively observe its effect on the accuracy of the simulation.

5.2.1. Adsorption data comparison

As introduced in Chapter 4, in order to run the simulation and to calculate the M-S surface diffusivities, adsorption data from 4 different sources were obtained from the literature. As it can be observed in Tab. 5.2 and Tab. 5.3, the adsorption data from publication 3 – obtained from GCMC simulations – consistently delivered the most inaccurate results especially in the extended Langmuir simulations. The simulations using the adsorption data from publications 1, 2, and 4 delivered roughly similar results, therefore supporting the consistency of the consulted data. However, an interesting observation from these data must be discussed. Considering diverse published literature, the data reported by publication 1 does not agree with the known preferential adsorption of ethane in ZIF-8. In addition when analyzing further Tab. 4.6, the ethene surface diffusivity calculated with this source is only 1.24 and 1.003 times faster than the ethane surface diffusivity in samples 3 and 4, respectively. These values result rather low when compared to the results obtained with the other publications (2.8 to 3.7 times faster). According to the work by Bux et al. [91], ethene should diffuse approximately 5 times faster than ethane at the pressure tested in this work, which supports the data obtained using publications 2, 3, and 4. Taking into account these different criteria, the analysis presented in the following sections was based on the results obtained with publication 2.

5.2.2. Counter-flux, feed composition, and permeance effect

Before any analysis is presented, several known behaviors of ZIF-8 must be introduced. First of all, when analyzing pure-component adsorption isotherms, the difference between ethane favored adsorption against ethene is not so significant, approximately 5-10% [91]. Nonetheless when speaking about the gas mixture, a known preferential adsorption of the paraffin over the olefin has been consistently reported in literature for so-called cation-free non-polar systems [91, 119, 120]. This is believed to be associated to a cooperation/competition effect between adsorbed molecules of like species. According to the experiments by Do and Do in similar systems [119], at a very low pressure range (0-500 Pa) a cooperation effect is consistently observed, exceeding the gas mixture adsorption isotherms those of the pure components. Then as the pressure is increased, the competition between species starts taking place and the isotherms of both gases are now lower when compared to their pure component isotherms, being the paraffin the stronger-adsorbing species. Furthermore, the reason behind such behavior might be associated to the molecular interactions between the methyl groups in ZIF-8 and the methyl group in ethane [120].

Relating this to the studied system, it is therefore very important to consider that the adsorption data from the single component isotherms will definitely not match exactly the conditions of the conducted experiments. In the case of the extended Langmuir model, single component adsorption data is used for all the estimations, for which the inaccuracies in the simulation results can be explained. On the other hand for the IAST simulations, a difference in the adsorption behavior between the single and binary mixture is somehow considered. For instance, in the work by Bux et al. [91], a very good description of the GCMC-IRM mixture isotherms when compared with IAST was reported, even in feed composition scenarios similar to the ones tested in this work. In Bux's work [91] it is also claimed that in contrast to the adsorption behavior, the loading dependent transport diffusion coefficients do not vary from single component to gas mixture, nor between feed composition scenarios.

To start with the analysis of the counter-flux effect it might be recalled as stated in section 5.1.1, that for the sake of simplicity and since both single and binary feed gas experiments were conducted with the same method, the counter-flux effect was considered to be indirectly incorporated via the experimental data collected rather than explicitly in the simulation. In order to verify the accuracy of this assumption, the counter-fluxes in the different experiments conducted were compared and presented in Tab. 5.4. As it can be seen, though both single and binary feed gas experiments presented a slight back permeation, a higher percentage was indeed observable in the single feed gas permeation experiments particularly for sample 3. Relating this to the assumption previously made, a counter-flux existed in both measurements – single and binary feed –, but the effect was not equal. Given this discrepancy in experimental conditions between the single and binary feed gas experiments, the estimated diffusion coefficients are therefore believed to be underestimated for the gas separation experiments thus influencing the simulation results.

Single feed gas scenario	Counter-flux (He) %	
	Sample 3	Sample 4
Pure ethene	22%	16%
Pure ethane	26%	18%
Binary feed gas scenario	**Sample 3**	**Sample 4**
Equimolar	12%	11%
Ethene-rich	16%	11%
Ethane-rich	13%	12%

Table 5.4: Counter-flux percentages in the single and binary feed gas experiments conducted on samples 3 and 4

Considering the diffusion coefficients from single feed gas permeation experiments might be underestimated, the diffusivities were tuned in the simulation program in order to observe the impact on the results. This was done on a trial and error basis, by increasing manually in the input parameters the M-S surface diffusivities for both gases, until the simulation fluxes matched with the experimental ones. This process was done for every membrane and every feed composition in order to obtain an average diffusivity since the diffusivities defined in the context of the M-S equations should not depend greatly on the mixture composition, i.e., should be constant with different experimental conditions [77]. As mentioned previously, all further simulation runs were done using the adsorption data from publication 2 [86].

As suspected, as the surface diffusivities were increased to the values presented in Tab. 5.5, the accuracy in the simulation significantly improved. Furthermore, it can be observed that the diffusivities, as expected, were quite constant within scenarios and that the estimated ratios between the ethene and ethane diffusivities are still within the expected ranges.

The average M-S surface diffusivity values, were then tested once again in all of the different scenarios for sample 3 and 4 for both adsorption models. The results are shown in Tab. 5.6. As it can be observed, after adjusting exclusively the M-S surface diffusivities in the simulation input parameters, the results for the IAST adsorption model delivered accurate results in all of the scenarios. On the other hand, the extended Langmuir delivered now considerably overestimated results as expected. To explain this, in order for the extended Langmuir model to be thermodynamically consistent, the q_{sat} values of all components should be the same [133, 134, 135]. Despite ignoring this unrealistic condition, this model has been proven over the years to be one of the most useful ones in describing mixed gas adsorption [133, 134, 135]. Nonetheless in this case, the mixture data does vary greatly from the single component, particularly regarding the preferential adsorption of ethane over ethene. It could then be concluded, that the results obtained in the first simulation runs for the extended Langmuir model were probably just coincidental since neither the counter-flux effect, nor the difference in adsorption data between single and binary gas mixtures were considered. To improve the simulation results in this model, the determination of mixed state parameters, e.g., saturation capacities and adsorption equilibrium constants, might be required in order for these specific interactions to be accounted for.

		M-S surface diffusivities $[m^2/s]$				
			Simulation (tuned)			
		Experimental	Equimolar	Ethene-rich	Ethane-rich	Average
Sample 3	Ethene	$4.86 \cdot 10^{-12}$	$10.12 \cdot 10^{-12}$	$9.39 \cdot 10^{-12}$	$9.75 \cdot 10^{-12}$	$9.75 \cdot 10^{-12}$
	Ethane	$1.31 \cdot 10^{-12}$	$2.40 \cdot 10^{-12}$	$2.07 \cdot 10^{-12}$	$2.40 \cdot 10^{-12}$	$2.29 \cdot 10^{-12}$
	Diffusivities ratio (Ethene/Ethane)	3.7	4.2	4.5	4.1	4.3
Sample 4	Ethene	$1.42 \cdot 10^{-12}$	$2.81 \cdot 10^{-12}$	$2.54 \cdot 10^{-12}$	$2.46 \cdot 10^{-12}$	$2.6 \cdot 10^{-12}$
	Ethane	$0.47 \cdot 10^{-12}$	$0.67 \cdot 10^{-12}$	$0.49 \cdot 10^{-12}$	$0.66 \cdot 10^{-12}$	$0.61 \cdot 10^{-12}$
	Diffusivities ratio (Ethene/Ethane)	3.0	4.2	5.2	3.7	4.3

Table 5.5: Experimental M-S surface diffusivities obtained from single feed gas permeation experiments and M-S surface diffusivities tuned in the simulation program for each of the scenarios tested

Finally, the difference between permeance values (i.e., presumed amount of defects) between samples 3 and 4 were compared using Tab. 5.6. A slightly higher accuracy could be observed in the results for sample 3. Also the ethane simulated fluxes were always slightly more inaccurate than the ethene simulated fluxes, and the most inaccurate result was consistently delivered for the ethane flux in the ethene-rich scenario. Reasons for such differences with membranes can be attributed to the several simplifications made during the simulation, including not considering explicitly the counter-flux effect, neglecting the contribution of micro-defects, and using literature adsorption data which also impacts in the accuracy of the estimated diffusivities.

In conclusion, after proper consideration of diverse factors, it could be said that the M-S surface diffusion model with a chemical potential gradient described via the IAST adsorption model provides an appropriate description of the binary gas mixture transport through the ZIF-8 SURMOF membrane. On the other hand, the extended Langmuir model being more mathematically simple, was not able to provide accurate results. This can be most likely attributed to the known cooperation/competition effect between the paraffins and olefins in so-called cation-free non-polar systems which experience a significant change in their adsorption behavior when compared to their pure-component isotherms [91, 119, 120].

Sample 3		Equimolar		Ethene-rich		Ethane-rich	
		J_1	J_2	J_1	J_2	J_1	J_2
IAST	Experimental	7.44	3.73	9.40	2.17	4.64	4.93
IAST	Simulated (Average)	7.17	3.56	9.77	2.41	4.63	4.72
IAST	Percent variation $\frac{J^{s(sim)}}{J^{s(exp)}}-1$	-4%	-5%	4%	11%	-0.2%	-4%
extended Langmuir	Experimental	7.44	3.73	9.40	2.17	4.64	4.93
extended Langmuir	Simulated (Average)	13.10	6.89	18.20	4.66	8.25	9.05
extended Langmuir	Percent variation $\frac{J^{s(sim)}}{J^{s(exp)}}-1$	76%	85%	94%	115%	78%	84%
Sample 4		**Equimolar**		**Ethene-rich**		**Ethane-rich**	
		J_1	J_2	J_1	J_2	J_1	J_2
IAST	Experimental	2.18	1.06	2.67	0.52	1.24	1.37
IAST	Simulated (Average)	2.02	0.96	2.75	0.65	1.30	1.28
IAST	Percent variation $\frac{J^{s(sim)}}{J^{s(exp)}}-1$	-7%	-9%	3%	25%	5%	-7%
extended Langmuir	Experimental	2.18	1.06	2.67	0.52	1.24	1.37
extended Langmuir	Simulated (Average)	3.74	1.88	5.19	1.27	2.36	2.48
extended Langmuir	Percent variation $\frac{J^{s(sim)}}{J^{s(exp)}}-1$	72%	77%	94%	144%	90%	81%

Table 5.6: Samples 3 and 4 simulated and experimental molar fluxes, J, $[\text{mmol}\cdot m^2\cdot s^{-1}]$ (after tuning) with their corresponding percent variation for different feed scenarios where 1 is ethene and 2 is ethane

Though this simulation was able to explain roughly the behavior of the simple system studied in this work, a considerable area of opportunity still exists regarding the modeling of MOF membrane-based separation. The achievement of more data is still required in order to improve several input parameters, for instance adsorption data (single and mixture) and coverage-dependent self-diffusivities which would aid in the description of the component interactions. Furthermore, different criteria when compared to already established models should be also considered given the particular properties MOFs possess, for instance the flexibility of their frameworks. Incorporating the defects and the fact that all of the pores might not be aligned in the same direction are also important points that must be accounted for in order to obtain a better description of the transport in these materials [136].

5.2.3. Single-file diffusion

Additionally, the single-file diffusion mode was also tested [78, 131]. This model considers a single narrow pore where there is only room for one type of molecule at any given time, and therefore the slowest moving molecule determines the permeance of all, i.e., there is no overtaking in the pore [72]. So Eq (1.14) reduces to:

$$J_i^s = -\frac{q_{sat} \cdot \rho \cdot D_i^s}{1 - \theta_i - \theta_j} \cdot ((1 - \theta_j)\nabla\theta_i + \theta_i \nabla\theta_j) \tag{5.3}$$

This model considers a one-dimensional motion of the molecules through the pores given their restricted mobility, i.e., narrow pore size [137]. Given the structure of ZIF-8 composed of four- and six-membered rings (approx. 3.4 Å in diameter) enclosing a larger cavity (approx. 11.1 Å in diameter), such diffusion mechanism wouldn't agree with theory considering ZIF-8's structure. Nonetheless, runs were conducted to confirm whether the performance of the membranes could be explained by assuming single-file diffusion. Similar to the previous section, the estimated average surface diffusivities and the adsorption data from publication 2 [86] were used in these simulations. The results are reported in Tab. 5.7. The poorly simulated molar fluxes confirmed that the molecule-molecule interactions in the adsorbed state, as well as the friction experienced by the molecules from the vacant sites do play an important role in the ZIF-8 SURMOF system. Therefore, they may not be neglected if an accurate description of the diffusion behavior in the system is sought.

	Experimental		Sample 3		Sample 4	
	$J_1^{s(exp)}$	$J_2^{s(exp)}$	$J_1^{s(sim)}$	$J_2^{s(sim)}$	$J_1^{s(sim)}$	$J_2^{s(sim)}$
Equimolar	7.44	3.73	0.037	-0.036	0.007	-0.002
Ethene-rich	9.40	2.17	0.075	-0.025	0.007	-0.003
Ethane-rich	4.64	4.93	0.026	-0.025	0.004	-0.001

Table 5.7: Simulated molar fluxes [mmol·m^2·s^{-1}] via the single-file diffusion mode for different scenarios for samples 3 and 4

5.2.4. Analyzing different number of discretization layers

Another point that was studied in this chapter, corresponds to the minimum number of discretization layers required to accurately model the transport behavior in the system. In the previous section, the parameter corresponding to the number of discretized layers, n_l, was set to 8; this sought to ensure that an accurate profile in each scenario was achieved regardless of the simulated profiles. In this section, this parameter was further analyzed in order to study a bit deeper the molar fraction and coverage profiles in each of the aforementioned scenarios. This was done by selecting the best performing adsorption model and adsorption data. In total, 4 runs were conducted for each scenario with different number of discretization layers. This process was conducted on both samples 3 and 4.

Considering the IAST adsorption model and using the adsorption data from publication 2 [86], the equimolar, ethene-rich, and ethane-rich feed composition scenarios were tested for different number of discretization layers: 1, 2, 4, and 8.

The data was collected and plotted as shown in Fig. 5.3 and Fig. 5.4 corresponding to the equimolar scenario profiles for samples 3 and 4, respectively. Each of these figures contains two graphs, one for the profiles of component 1, i.e., ethene, and the second one for the profiles of component 2, i.e., ethane. In each graph the molar fraction and surface coverage profiles are simultaneously depicted starting with the interface between the retentate chamber and the membrane surface and ending in the interface between the ZIF-8 SURMOF layer and the α-Al_2O_3 support.

The graphs mentioned in the previous paragraph, as well as for the ethene-rich and ethane-rich scenarios, showed that the molar fraction profiles followed a rather linear behavior. Such conclusion could be reached as increasing the number of discretization layers (up to n_l=8) resulted in no major changes or alterations in the simulated molar fraction profiles. The rest of the graphs plotted for the different scenarios can be found in Appendix A.6. Regarding the surface coverage profiles, a linear behavior could also be observed. For $n_l \geq 2$ no major improvement in the simulated profiles could be noticed. This point was further studied using the previously fitted adsorption data presented in Chapter 4 (adsorption constants and saturation capacities). The surface coverages for ethene and ethane in ZIF-8 were then estimated, supporting that at the pressure range studied in this work, the coverage falls indeed within the linear section of its profile, i.e., the simulation results fit the expected behavior. Important and interesting to mention is that as the pressure starts to increase, the adsorption behavior is no longer lineal and the difference between the ethene and ethane adsorption at higher pressures (>6 bar) appears to be reduced, for which operating at higher pressures could also prove beneficial provided the membrane stability is not compromised keeping in mind the mixture behavior could of course differ. In the case of the presented work, it may be concluded that at P_{abs} =1.1 bar a discretization of 2 layers, i.e., n_l=2, is enough to accurately model both, the molar fractions and surface coverage profiles through the ZIF-8 SURMOF pores. If a higher absolute pressure were to be tested ($\geq$2.5 bar), a greater number of discretization layers would be indeed required.

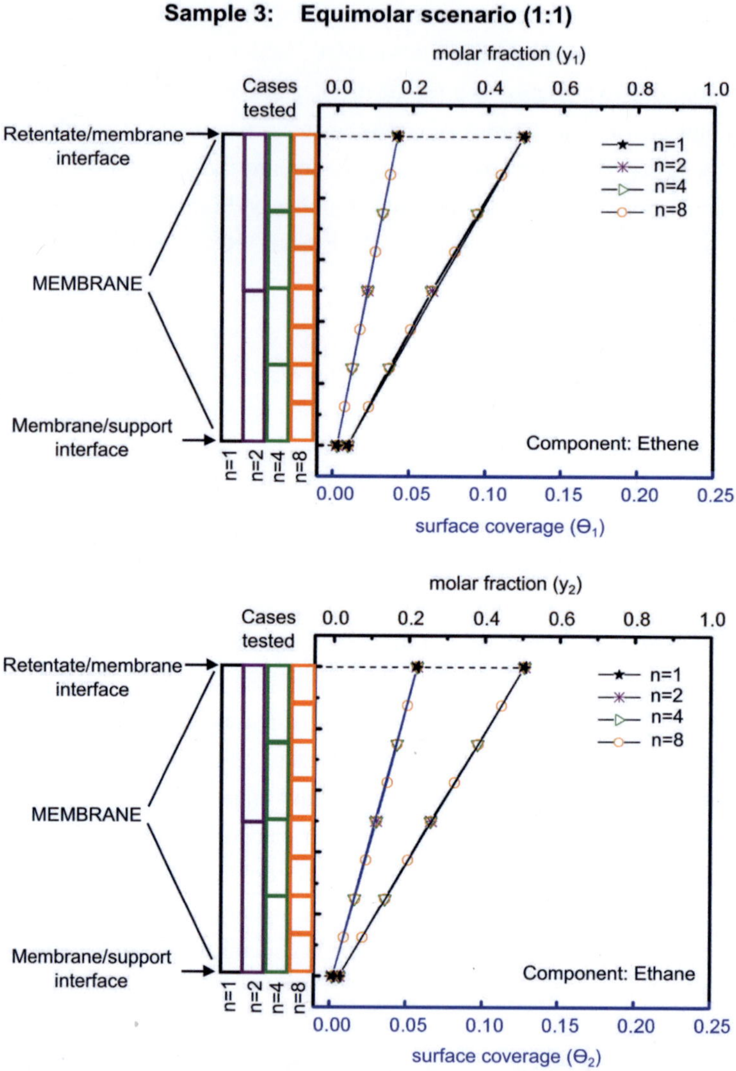

Figure 5.3: Simulated y and θ profiles (IAST adsorption model) for different number of discretization layers of sample 3 under equimolar feed

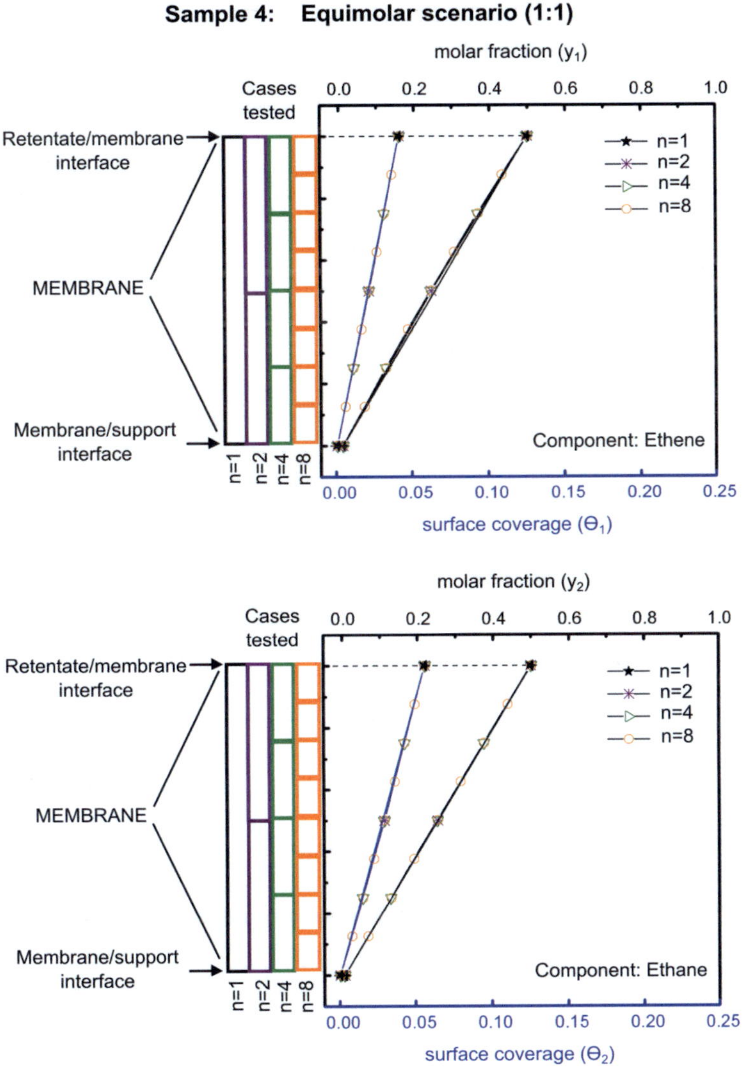

Figure 5.4: Simulated y and θ profiles (IAST adsorption model) for different number of discretization layers of sample 4 under equimolar feed

6. Summary and Outlook

The LPE LBL ZIF-8 SURMOF synthesis for application in gas separation was explored thoroughly in this work. To begin in Chapter 2, non-coated and Au-coated α-Al_2O_3 supports were studied and contrasted with diverse characterization techniques including XRD, IRRAS, and SEM. The study regarding the number of deposition cycles showed that although the synthesis was successful on both supports, the Au-coated supports delivered better homogeneous crystals in terms of size and roughness. Also the Au-coated supports presented higher predominant orientation in the (110) plane, which is essential in order to offer the best access to guest molecules during the gas separation. Furthermore, the minimum number of deposition cycles in order to achieve a full coverage of the Au-coated surface was also studied via H_2 pressure-dependent experiments and was determined to be 175 deposition cycles. Defining this parameter was of great importance in order to ensure an overall high membrane quality, though the presence of micro-defects are not distinguishable with this test.

After the synthesis of the ZIF-8 SURMOF membranes was optimized, the thermal and mechanical stability of the crystals was also studied in Chapter 3. By ensuring the crystals are stable to relevant conditions, accurate experimental data was achieved and thus a reliable simulation study could be conducted in Chapter 5. After stability treatment and compression experiments, it was determined that the established LPE LBL ZIF-8 SURMOF system is stable under exposure of ethene and ethane at normal temperature and pressure, showing no change in crystallinity, morphology, or composition. Furthermore, the temperature limit was found to be around 100 °C for both gases. Nonetheless, it was interesting to observe that the crystals appear more stable under ethane than ethene up to 150 °C. However, as the temperature was increased above this point, the effect for ethane intensified and completely destroyed the crystallinity and composition of the ZIF-8 crystals, while the ethene-exposed crystals still showed some broadened IR bands. Regarding the stability of the crystals under various pressures, the crystallinity remained intact even after a 100 N force ($P_{abs} \approx 8.5$ bar) was applied. Nevertheless, though the device used to compress the membranes permits the evaluation of any immediate change in crystallinity, it does not help identify the limit of the system regarding crack formations. The force applied with this device is not the same type of force applied during the gas separation tests, i.e., it does not ensure an even pressure over the surface of the membrane. Still, based on the SEM images for the 100 N scenario, no cracks were visibly observable. Also, considering performed permeation and gas separation experiments, the membranes consistently delivered the same results, even within measurement spans of several months and measurement times involving 10 continuous hours, suggesting membrane stability at least at $P_{abs} = 1.5$ bar or below.

In Chapter 4, samples 1, 2, 3, and 4 corresponding to LPE LBL ZIF-8 SURMOF membranes deposited for 175 cycles, were tested for binary feed gas separation of ethene and ethane using the Wicke-Kallenbach method. Different feed compositions and pressure ranges were studied. Samples 1 and 2 delivered low ethene selectivity with C_2H_4/C_2H_6 separation factors (SFs) ranging between 1.25 and 1.46, while samples 3 and 4 exhibited better separation reaching C_2H_4/C_2H_6 SFs of up to 2.60. These last two membranes also showed different performance depending on the feed composition, suggesting the presence of minimal grain boundary defects reflected via selective transport. These results also agree with the lower obtained H_2 permeances in comparison to samples 1 and 2 as reported in Chapter 2. The pressure range tested showed very slight effects on the separation factors which decreased with increasing pressure in most cases; such effect is believed to be associated with induced rotational displacement of the imidazole ligands with pressure variations. The counter-flux of the sweep gas given the method used was also analyzed, suggesting no great effect in samples 3 and 4 during the binary feed gas separation experiments.

Considering the performance results, samples 3 and 4 were further studied in single feed gas permeation experiments to obtain experimental data required for the simulation study of this work. In order to obtain consistent parameters, the single feed gas permeation experiments were also performed using the Wicke-Kallenbach method. This was of particular importance since for the sake of simplicity, the counter-flux effect during the gas separation was not considered directly in the simulation. Instead, it was considered implicitly via the experimental data, at least as a first approximation, therefore the importance of consistency within experiments. Following, based on the measurements and on diverse literature, the surface coverage on the retentate and permeate side were calculated in order to further estimate the M-S surface diffusivities in each of these membranes. Combining then the single feed gas experimental data with various adsorption data of ZIF-8 SURMOF crystals found in the literature, the M-S surface diffusivities for samples 3 and 4 were estimated.

As the last step of this work, the experimental data was used to understand the diffusion behavior through the synthesized ZIF-8 SURMOF membranes and the role different parameters play in it. To do so, a MATLAB®-simulation program, based on the generalized M-S surface diffusion equations by Krishna and co-workers, was implemented from the works by Ding and adapted to the specific needs of this system. Two different adsorption models were studied, IAST and extended Langmuir, four different sources for adsorption data were used, and three different scenarios were tested. First of all, the consistency of the consulted adsorption data was studied with the obtained results and the most appropriate source was selected for further simulations. Following, the two different adsorption models were compared.

From the preliminary results, the simulation implemented with the extended Langmuir model seem to deliver better results when compared to the IAST model, but in order to reach further conclusions several points were revised. First of all, the counter-flux of the sweep gas to the retentate chamber from the permeate chamber was confirmed to variate somewhat within experiments specially in the single feed gas permeation experiments, delivering underestimated surface diffusivities. To improve the approximation of the counter-flux effect, the diffusivities were tuned in the simulation on a trial and error basis for each membrane and scenario, and an average M-S surface diffusivity for each gas was obtained considering that on the context of the M-S equations, the diffusivities should be constant with different experimental conditions. After these considerations, the simulation results with the IAST model delivered accurate results, whereas the extended Langmuir model delivered overestimated fluxes agreeing with the adsorption behavior differences between the single and binary gas mixtures.

Finally, the minimum number of discretization layers, n_l, was also analyzed. Based on diverse runs with different n_l (1, 2, 4, and 8), it could be concluded that the simulated molar fractions and surface coverage profiles possess a rather linear behavior. Using the fitted adsorption data presented in Chapter 4, the surface coverage for ethene and ethane were estimated, falling definitely within the linear section of their corresponding profiles and confirming that the simulation results fit the expected behavior. Interesting though, is that as the pressure starts to increase, the adsorption behavior is no longer lineal and the difference between the ethene and ethane adsorption at higher pressures ($>$6 bar) appears to be reduced, for which operating at higher pressures could also prove beneficial provided the membrane stability is not compromised and keeping in mind that the mixture behavior could of course differ. It could then be concluded that for the ZIF-8 SURMOF system studied in this work, 2 discretization layers are enough to approximate its transport behavior. If higher absolute pressures were to be tested ($\geq$2.5 bar), a greater number of discretization layers would indeed be required.

To conclude, this work was able to successfully synthesize LPE LBL ZIF-8 SURMOF membranes on Au-coated α-Al_2O_3 supports with an overall promising quality in terms of crystallinity and orientation, i.e., confirming the liquid phase epitaxial layer-by-layer approach as a good alternative for MOF synthesis without the use of high temperatures or pressures. The ethene-selective membranes reached SFs of up 2.6 depending on the feed composition, which was in good agreement with other reported synthesis techniques, but still away from feasible industrial application. Given the preferential adsorption of ethane competing against the preferential diffusion of ethene, achieving high separation factors is quite challenging even within highly oriented ZIF-8 crystals, for which further tailoring of the current system could prove helpful, e.g., post-synthesis methods, customized linker molecules, etc.

Understanding the transport within the ZIF-8 pores is also a critical step in the design of such target-wise systems. For the sake of simplicity several assumptions were made for the simulation implemented in this work. Though a good insight and approximate description was achieved, still some inaccuracies could be clearly observed particularly when comparing the different samples studied in this work. To achieve greater knowledge on the description of this and other MOFs systems, more data is still required. First of all, more membranes must be tested to have a larger amount of performance data, including their behaviors with feed compositions as well as the degree of reproducibility of the synthesis technique.

Second, the determination of the micro-defects on the membranes is of great importance for several reasons. On the one hand, this would permit to improve the simulation by considering the diffusion through such micro-defects according to their pore-size which should definitely aid in the accuracy of the separation description. On the other hand, having the knowledge of the exact quality of the membrane could also support the qualitatively identification of trends regarding the adsorption behavior. The determination of micro-defects via permporometry techniques is one of the main points to be continued in next works.

Another important area of improvement is related to the adsorption data. Though diverse literature was consulted for this work, the achievement of adsorption data from the LPE LBL ZIF-8 SURMOF synthesized membranes would be quite appropriate to achieve precise input parameters for the simulation. For this, a langatate crystal microbalance can be used, which additionally offers the possibility of acquiring pure-component adsorption data at higher temperatures and pressures. This would provide further information of the ZIF-8 crystals, for instance whether operating at considerably higher pressures could also help in the separation performance and perhaps open the doors for further application possibilities. Ideally, the adsorption data of a binary system should be considered, particularly for ZIF-8 in which the pure-component adsorption data appears to vary greatly from the pure-component one, but the experimental path to achieve such data is still to be reflected upon.

Moreover the diffusivity parameters must be also revised and improved. Regarding the M-S surface diffusivities, the use of precise adsorption data would be a first clear point of improvement. Performing single gas permeation experiments without the use of sweep gas could also be quite advantageous in order to remove any counter-flux effect. With respect to the cross-term diffusivity a full formalism is yet to be achieved, i.e., by experimentally obtaining first the self-diffusivities. As mentioned in the introduction, the cross-term diffusivity is rather difficult to obtain experimentally, but efforts in this direction should not be discarded.

Finally, when speaking about simulation of MOF membrane-based membrane-based separation systems, the consideration of certain MOF features could also prove helpful including the framework flexibility and the existence of different pore alignments. With proper consideration and analysis of all these above points, the design of target-wise systems that can exceed actual MOF performances in this and other fields might be possible.

List of Figures

List of Tables

Abbreviations

as-syn.	as-synthesized
CMS	Carbon molecular sieves
comp.	components
eL	Extended Langmuir
GMS	Generalized Maxwell-Stefan surface diffusion model
IR	Infrared
IRRAS	Infrared reflection absorption spectroscopy
IUPAC	International Union of Applied Chemistry
LBL	Layer-by-layer
LPE	Liquid phase epitaxy
MD	Molecular dynamics
M-S	Maxwell-Stefan
MOF	Metal-organic framework
MUD	11-mercapto-1-undecanol
PFG-NMR	Nuclear magnetic resonance pulsed field gradient
PVD	Physical vapor deposition
QENS	Quasielastic neutron scattering
SAM	Self-assembled monolayer
SF	Separation factor
SOD	Sodalite
SURMOF	Surface anchored metal-organic framework
XRD	X-ray diffraction
ZIF	Zeolitic imidazolate frameworks

Nomenclature

α	Separation factor
∇C	Concentration gradient
d_p	Diameter of the pore
D^{BF}	Binary friction diffusivity
D^K	Knudsen diffusivity
D_{ij}^m	Binary diffusivity
D^s	M-S surface diffusivity
D_{ij}^s	M-S cross-term surface diffusivity
$D_{i,i}^s$	M-S self-exchange diffusivity
$D_{self,i}^s$	M-S self-diffusivity
δ_m	Membrane thickness
δ_s	Support thickness
ε_m	Membrane's porosity
ε_s	Support's porosity
η	Viscosity of the fluid
Γ	Thermodynamic factor
J^K	Knudsen molar flux
J^s	Surface molar flux
$J^{s(sim)}$	Simulated surface molar flux
$J^{s(exp)}$	Experimental surface molar flux
J^m	Molar flux by molecular diffusion
J^v	Viscous molar flux
k	Adsorption constant

K_n	Knudsen number
k_{sch}	Scherer's constant
λ_m	Mean free path
M	Molar mass
$\nabla\mu$	Chemical potential gradient
n	Number of molecules
n_l	Number of discretized layers
P	Pressure
p_i	Partial pressure
$p_i^0(\pi)$	Partial pressure of pure component at T and P of mixture
p_p	Partial pressure on permeate side
p_r	Partial pressure on retentate side
$\mathfrak{p}$	Single gas permeability
q_i^0	Standard state loading
q_{sat}	Saturation capacity
q_{tot}	Total loading of adsorbed molecules
R	Universal gas constant
r_p	Pore radius
ρ_m	Density of the membrane
σ	Diameter of the ideal gas spherical molecules
S_{ij}	Ideal selectivity or permselectivity
T	Temperature
τ_m	Tortuosity of the membrane
θ	Surface coverage
θ_p	Surface coverage on permeate side
θ_r	Surface coverage on retentate side
$\overline{\theta_i}$	Average coverage
$\nabla\theta_i$	Gradient of coverage
υ	Atomic diffusion volumes
x	Molar fraction in the adsorbed phase
y	Molar fraction in the gas phase

y_p	Molar fraction on the permeate side
y_r	Molar fraction on the retentate side

A. Appendix

A.1. ZIF-8 powder diffractrogram

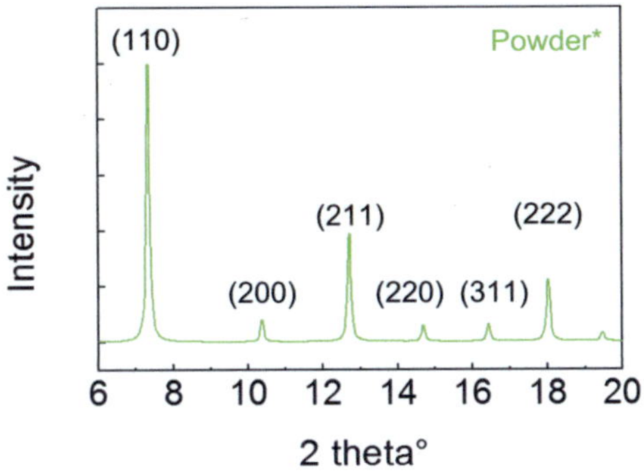

Figure A.1: Powder diffractogram of ZIF-8. Figure created by author from data in [46]

A.2. Substrate Description and Preparation

A.2.1. Permeation experiments

Permeation experiments were also additionally performed to confirm experimentally if the porosity of the support had been compromised with the Au thin film addition. Argon was used as the inlet flow with ranges between 17.5- 105 ml· min^{-1}. The permeate vs. feed percentages for increasing delta P's were recorded as shown in Tab. A.1. From this table, it can be confirmed that the thin Au film is not blocking the pores and that the coated supports presented surprisingly even higher permeating percentages, confirming these last ones are mainly attributed to the finite reproducibility of the α-Al_2O_3 support itself (Au film thickness $\approx$ 70 nm; α-Al_2O_3 support thickness= 1 mm), which as reported by Fraunhofer IKTS ranges between 40-55%.

	Permeating Flows %		
Delta P	Permeate / Feed	Permeate / Feed	Permeate / Feed
mbar	Support 1 α-Al2O3	Support 2 α-Al2O3 + Au	Support 3 α-Al2O3 + Au
4.1	10.11%	16.23%	12.69%
5.2	10.00%	16.15%	12.49%
6.3	10.34%	15.98%	11.91%
8.7	10.40%	16.91%	11.84%
11.3	10.92%	17.30%	11.75%
26.8	12.88%	19.19%	13.54%

Table A.1: Permeating percentages of Argon with increasing delta P's along non-coated and Au-coated α-Al_2O_3 supports (adapted from [35])

A.2.2. Pressure-dependent experiments on various supports

Different supports – non-coated and Au-coated – were also investigated via H_2 pressure dependent experiments. As it can be observed in Fig. A.2, the tests further confirmed that no two supports are exactly the same. Slight variations among synthesized membranes can be in part attributed to these differences observed in the supports. For further studies, a systematic study starting with non-coated α-Al_2O_3 to ZIF-8 SURMOF membranes is of interest to understand the importance and role every single step plays in the process.

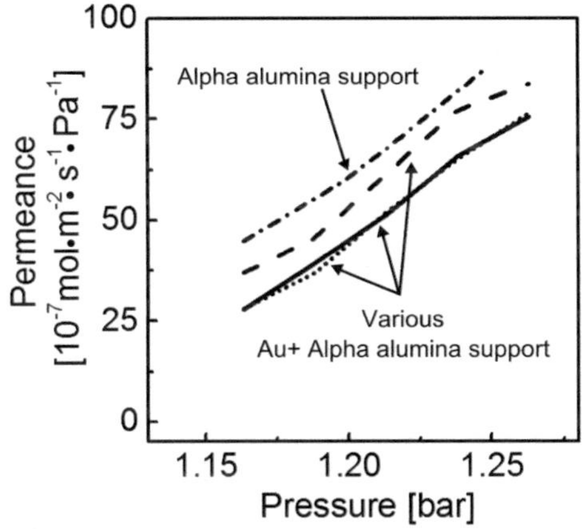

Figure A.2: H_2 permeance through various supports (coated and non-coated) tested for increasing feed absolute pressures at room temperature

A.3. Growth study on various porous α-Al_2O_3 supports

As described in the introduction, the achievement of highly oriented ZIF-8 SURMOF crystals on the (110) plane is of high importance in order to provide a facilitated access to guest molecules. In Chapter 2, it was shown how the synthesized ZIF-8 SURMOF via the LPE LBL technique presented a predominant orientation in the (110) plane. As described later on, the Au-coated supports resulted more oriented presenting higher percentage of intensity reduction on the other planes when compared to the ZIF-8 powder. In Fig. A.3 the percentages (increment or reduction) are shown in detail for both supports for 200 deposition cycles experiments. As stated previously, the XRD data for the powder ZIF-8 was extracted from the published work by Lewis [46]. Furthermore, using the Diffrac.EVA V2.1 program from Bruker with a Scherer constant of k_{sch}=0.89, the integral breadths were also obtained and shown in Tab. A.2. As it can be seen, the lower values reported correspond to the (110) and (220) planes, further confirming a larger crystallite size domain on these planes.

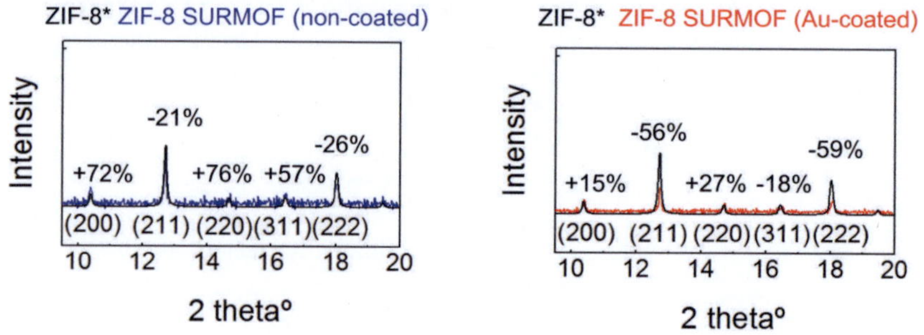

Figure A.3: Normalized intensity values and percentages changes for ZIF-8 powder [46] and ZIF-8 SURMOF on non-coated and Au-coated supports

	Integral breadth	
Plane	ZIF-8 SURMOF (non-coated)	ZIF-8 SURMOF (Au-coated)
(110)	0.093	0.078
(200)	0.117	0.088
(211)	0.116	0.104
(220)	0.087	0.081
(311)	0.116	0.163
(222)	0.115	0.114

Table A.2: Integral breadths for ZIF-8 SURMOF deposition on non-coated and Au-coated α-Al_2O_3 supports

A.4. Stability treatments

A.4.1. Williamson and Hall Plots

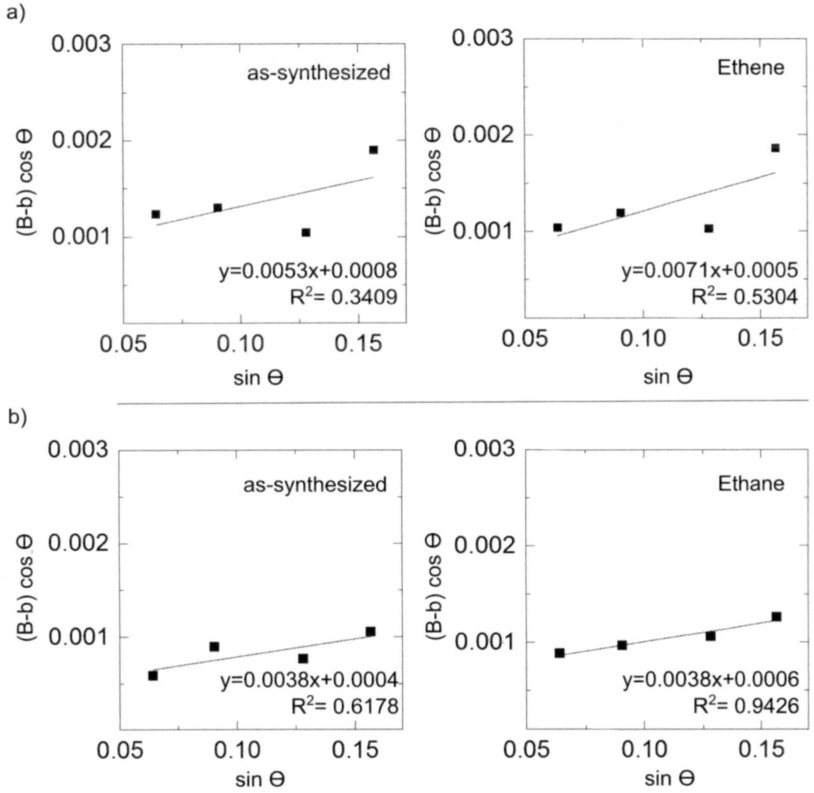

Figure A.4: Williamson and Hall plots for as-synthesized and a) ethene- or b) ethane-exposed ZIF-8 SURMOF membranes at room temperature

A.4.2. XRD patterns for stability treatments at 200 °C

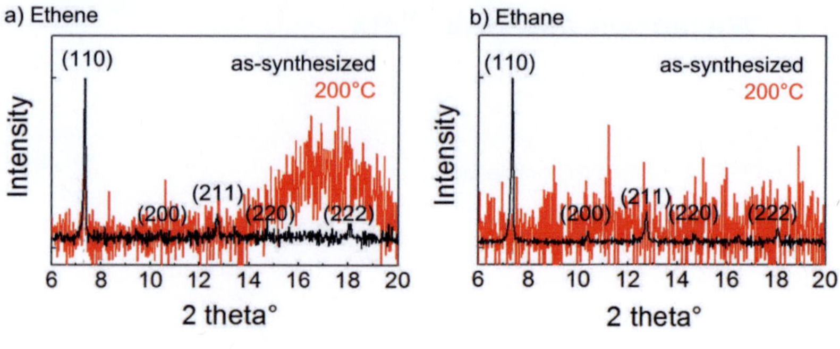

Figure A.5: Normalized XRD patterns of ZIF-8 SURMOF membranes exposed to a constant stream of a) ethene or b) ethane for 24 h

A.5. Permeance through ZIF-8 SURMOF membranes (pressure variation)

As mentioned in Chapter 4, the gas separation experiments were conducted at an absolute pressure range of 1.1-1.3 bar seeking to observe whether an effect in the performance existed even with small pressure variations. No relevant effect was observable at such pressure range, for which all further experiments were conducted at 1.1 bar. In the next pages, Fig. A.6, A.7, A.8, and A.9 show the ethane and ethene permeance, as well as the separation factors as a function of pressure for the different samples tested.

A.5.1. Sample 1

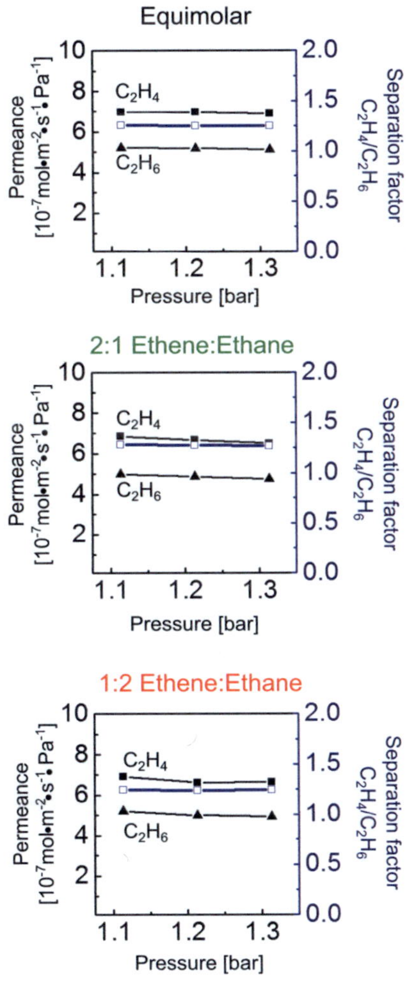

Figure A.6: Ethene permeance, ethane permeance, and C_2H_4/C_2H_6 separation factors of sample 1 (ZIF-8 SURMOF) for different feed compositions

A.5.2. Sample 2

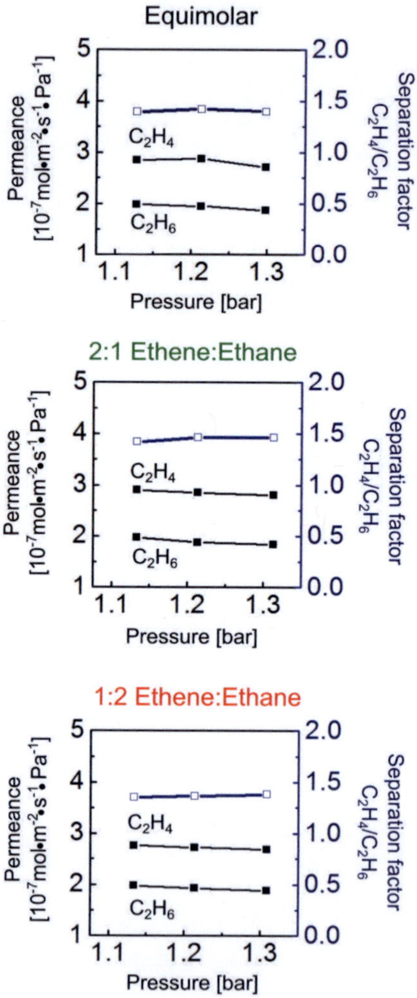

Figure A.7: Ethene permeance, ethane permeance, and C_2H_4/C_2H_6 separation factors of sample 2 (ZIF-8 SURMOF) for different feed compositions

A.5.3. Sample 3

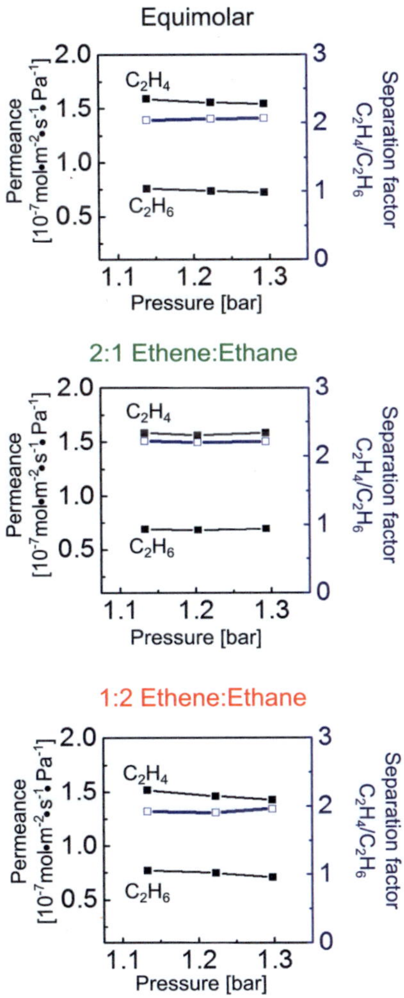

Figure A.8: Ethene permeance, ethane permeance, and C_2H_4/C_2H_6 separation factors of sample 3 (ZIF-8 SURMOF) for different feed compositions

A.5.4. Sample 4

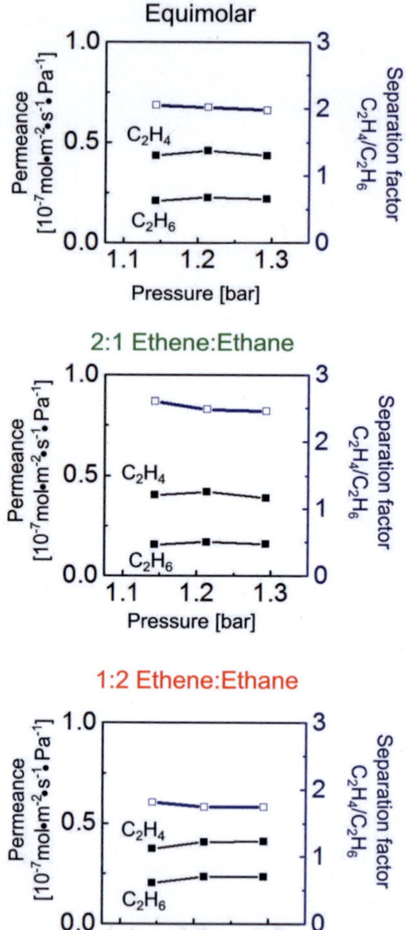

Figure A.9: Ethene permeance, ethane permeance, and C_2H_4/C_2H_6 separation factors of sample 4 (ZIF-8 SURMOF) for different feed compositions

A.6. Simulated y and θ profiles

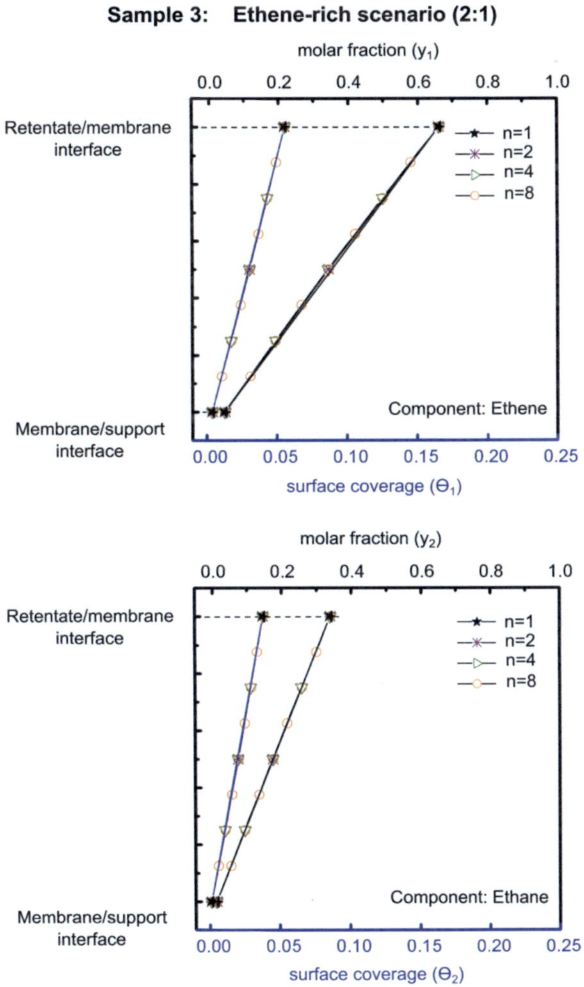

Figure A.10: Simulated y and θ profiles (IAST adsorption model) for different number of discretization layers of sample 3 under ethene-rich feed

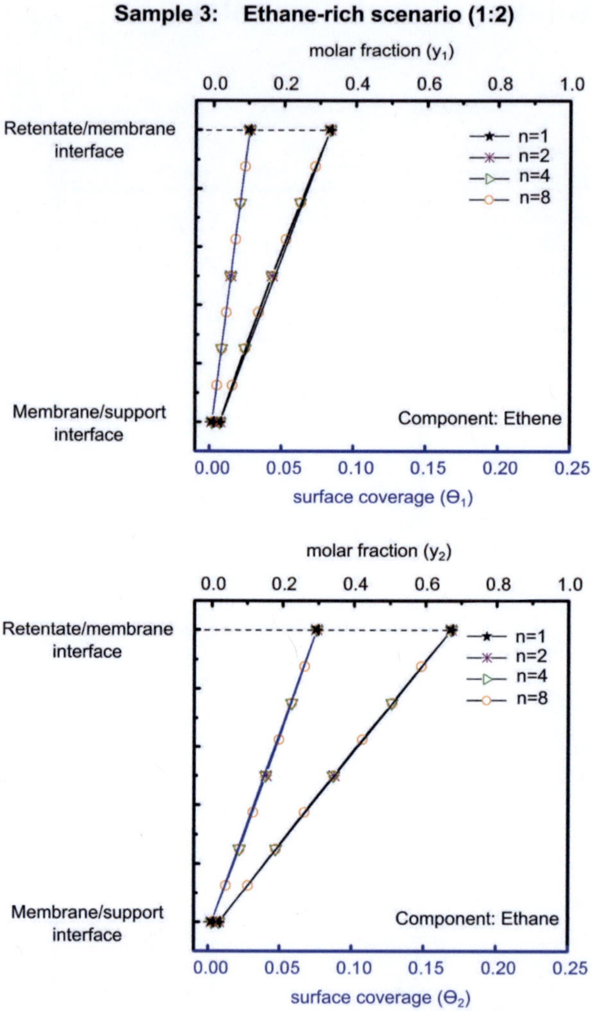

Figure A.11: Simulated y and θ profiles (IAST adsorption model) for different number of discretization layers of sample 3 under ethane-rich feed

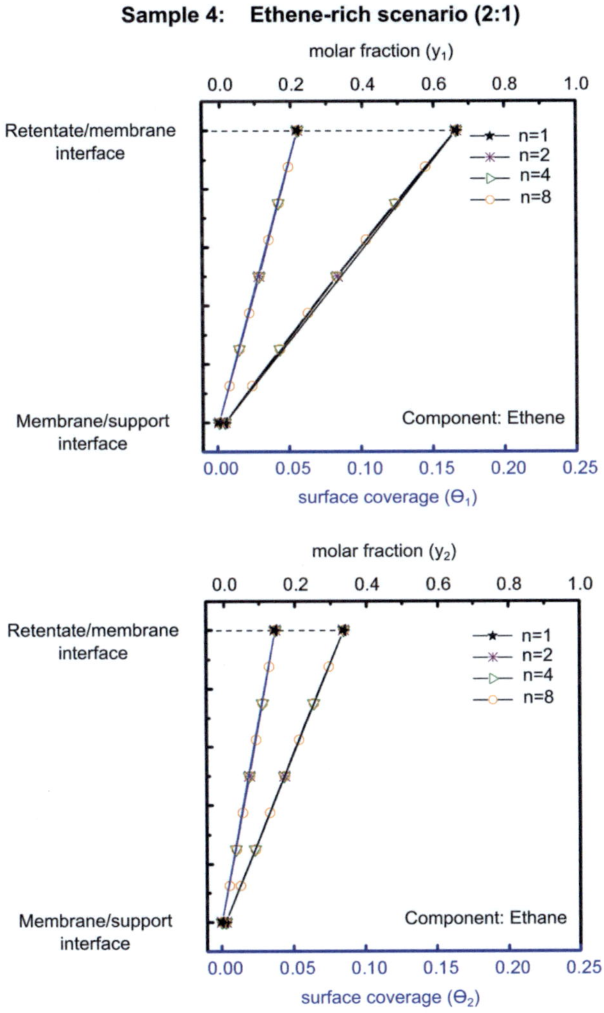

Figure A.12: Simulated y and θ profiles (IAST adsorption model) for different number of discretization layers of sample 4 under ethene-rich feed

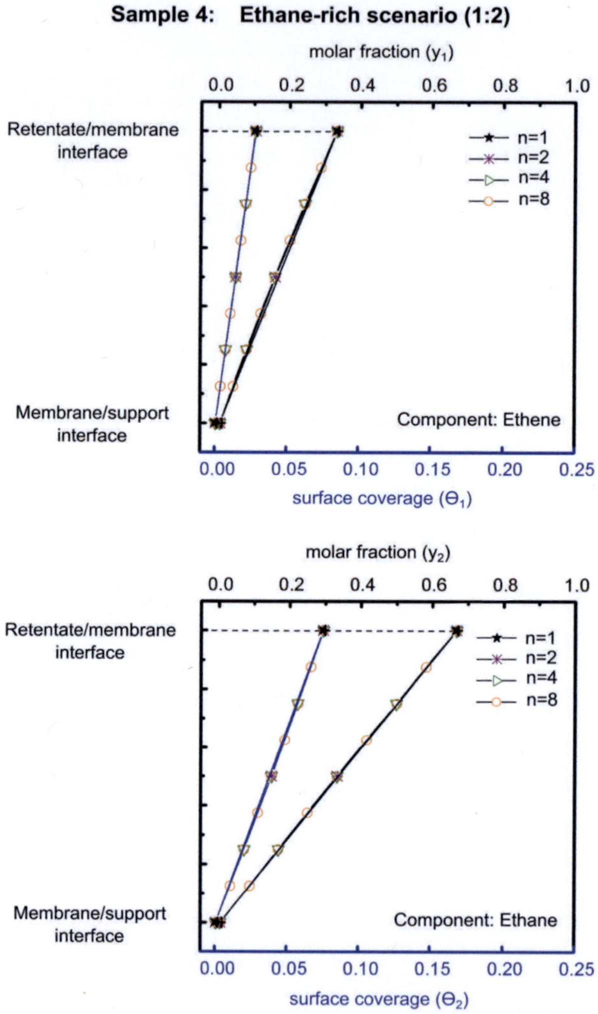

Figure A.13: Simulated y and θ profiles (IAST adsorption model) for different number of discretization layers of sample 4 under ethane-rich feed

References

[1] Rami Faiz and Kang Li. Olefin/paraffin separation using membrane based facilitated transport/chemical absorption techniques. *Chemical Engineering Science*, 73:261–284, 2012.

[2] João G Crespo and Karl W Böddeker. *Membrane processes in separation and purification*, volume 272. Springer Science & Business Media, 1994.

[3] Anil K Pabby, Syed SH Rizvi, and Ana MS Requena. *Handbook of membrane separations: chemical, pharmaceutical, food, and biotechnological applications*. CRC press, 2008.

[4] Chen Zhang. Alkane and alkene separation by membrane operations. In *Encyclopedia of Membranes*, pages 55–56. Springer, 2016.

[5] Ahmad F Ismail, Kailash C Khulbe, and Takeshi Matsuura. *Gas Separation Membranes*. Springer, 2015.

[6] Lloyd M Robeson. Correlation of separation factor versus permeability for polymeric membranes. *Journal of membrane science*, 62(2):165–185, 1991.

[7] Nidal Hilal, Mohamed Khayet, and Chris J Wright. *Membrane modification: Technology and applications*. CRC press, 2012.

[8] Jorge Gascon, Freek Kapteijn, Beatriz Zornoza, Victor Sebastian, Clara Casado, and Joaquin Coronas. Practical approach to zeolitic membranes and

coatings: state of the art, opportunities, barriers, and future perspectives. *Chemistry of Materials*, 24(15):2829–2844, 2012.

[9] Jian-Rong Li, Julian Sculley, and Hong-Cai Zhou. Metal–organic frameworks for separations. *Chemical reviews*, 112(2):869–932, 2012.

[10] Aisheng Huang, Yifei Chen, Nanyi Wang, Zhongqiao Hu, Jianwen Jiang, and Jürgen Caro. A highly permeable and selective zeolitic imidazolate framework zif-95 membrane for h2/co2 separation. *Chemical Communications*, 48(89):10981–10983, 2012.

[11] Jianfeng Yao and Huanting Wang. Zeolitic imidazolate framework composite membranes and thin films: synthesis and applications. *Chemical Society Reviews*, 43(13):4470–4493, 2014.

[12] Osama Shekhah, Hui Wang, Stefan Kowarik, Frank Schreiber, Michael Paulus, Metin Tolan, Christian Sternemann, Florian Evers, Denise Zacher, Roland A Fischer, et al. Step-by-step route for the synthesis of metal-organic frameworks. *Journal of the American Chemical Society*, 129(49):15118–15119, 2007.

[13] Osama Shekhah, Raja Swaidan, Youssef Belmabkhout, Marike Du Plessis, Tia Jacobs, Leonard J Barbour, Ingo Pinnau, and Mohamed Eddaoudi. The liquid phase epitaxy approach for the successful construction of ultra-thin and defect-free zif-8 membranes: pure and mixed gas transport study. *Chemical Communications*, 50(17):2089–2092, 2014.

[14] Osama Shekhah. Layer-by-layer method for the synthesis and growth of surface mounted metal-organic frameworks (surmofs). *Materials*, 3(2):1302–1315, 2010.

[15] Jin-Liang Zhuang, Andreas Terfort, and Christof Wöll. Formation of oriented and patterned films of metal–organic frameworks by liquid phase epitaxy: A review. *Coordination Chemistry Reviews*, 307:391–424, 2016.

[16] Lars Heinke, Zhigang Gu, and Christof Wöll. The surface barrier phenomenon at the loading of metal-organic frameworks. *Nature communications*, 5:4562, 2014.

[17] Miral Shah, Michael C McCarthy, Sonny Sachdeva, Alexander K Lee, and Hae-Kwon Jeong. Current status of metal–organic framework membranes for gas separations: promises and challenges. *Industrial & Engineering Chemistry Research*, 51(5):2179–2199, 2012.

[18] Jürgen Caro. Are mof membranes better in gas separation than those made of zeolites? *Current Opinion in Chemical Engineering*, 1(1):77–83, 2011.

[19] Kyo S Park, Zheng Ni, Adrien P Côté, Jae Y Choi, Rudan Huang, Fernando J Uribe-Romo, Hee K Chae, Michael O'Keeffe, and Omar M Yaghi. Exceptional chemical and thermal stability of zeolitic imidazolate frameworks. *Proceedings of the National Academy of Sciences*, 103(27):10186–10191, 2006.

[20] Alexander U Czaja, Natalia Trukhan, and Ulrich Müller. Industrial applications of metal–organic frameworks. *Chemical Society Reviews*, 38(5):1284–1293, 2009.

[21] Omar M Yaghi, Guangming Li, and Hailian Li. Selective binding and removal of guests in a microporous metal–organic framework. *Nature*, 378(6558):703–706, 1995.

[22] Jinxuan Liu, Binit Lukose, Osama Shekhah, Hasan K Arslan, Peter Weidler, Hartmut Gliemann, Stefan Bräse, Sylvain Grosjean, Adelheid Godt, Xinliang Feng, Klaus Müllen, Ioan-Bogdan Magdau, Thomas Heine, and Christof Wöll. A novel series of isoreticular metal organic frameworks: realizing metastable structures by liquid phase epitaxy. *Scientific reports*, 2:921, 2012.

[23] Osama Shekhah and Mohamed Eddaoudi. The liquid phase epitaxy method for the construction of oriented zif-8 thin films with controlled growth on

functionalized surfaces. *Chemical Communications*, 49(86):10079–10081, 2013.

[24] Hartmut Gliemann and Christof Wöll. Epitaxially grown metal-organic frameworks. *Materials today*, 15(3):110–116, 2012.

[25] An-Hui Lu and Sheng Dai. *Porous materials for carbon dioxide capture*. Springer Science & Business, 2014.

[26] Sebastian Henke. *Metal-Organic Frameworks with Additional Flexible Substituents – Modulating Responsiveness, Gas Sorption Selectivity & Network Topologies*. PhD thesis, Ruhr-Universität Bochum, 2011.

[27] Ulrich Müller, Markus M Schubert, Friedhelm Teich, Hermann Puetter, Kerstin Schierle-Arndt, and Jörg Pastré. Metal–organic frameworks—prospective industrial applications. *Journal of Materials Chemistry*, 16(7):626–636, 2006.

[28] Ashlee J Howarth, Yangyang Liu, Peng Li, Zhanyong Li, Timothy C Wang, Joseph T Hupp, and Omar K Farha. Chemical, thermal and mechanical stabilities of metal–organic frameworks. *Nature Reviews Materials*, 1:15018, 2016.

[29] Shilun Qiu, Ming Xue, and Guangshan Zhu. Metal–organic framework membranes: from synthesis to separation application. *Chemical Society Reviews*, 43(16):6116–6140, 2014.

[30] Miral N Shah. *Synthesis and Characterization of Films and Membranes of Metal-Organic Framework (MOF) for Gas Separation Applications*. PhD thesis, Texas A&M University, 2012.

[31] Helge Bux, Armin Feldhoff, Janosch Cravillon, Michael Wiebcke, Yan-Shuo Li, and Jürgen Caro. Oriented zeolitic imidazolate framework-8 membrane with sharp h2/c3h8 molecular sieve separation. *Chemistry of Materials*, 23(8):2262–2269, 2011.

[32] Hasan K Arslan, Osama Shekhah, Jonas Wohlgemuth, Matthias Franzreb, Roland A Fischer, and Christof Wöll. High-throughput fabrication of uniform and homogenous mof coatings. *Advanced functional materials*, 21(22):4228–4231, 2011.

[33] Valeriya Chernikova, Osama Shekhah, and Mohamed Eddaoudi. Advanced fabrication method for the preparation of mof thin films: Liquid-phase epitaxy approach meets spin coating method. *ACS Applied Materials & Interfaces*, 8(31):20459–20464, 2016.

[34] Osama Shekhah, Hui Wang, Denise Zacher, Roland A Fischer, and Christof Wöll. Growth mechanism of metal–organic frameworks: Insights into the nucleation by employing a step-by-step route. *Angewandte Chemie International Edition*, 48(27):5038–5041, 2009.

[35] Elvia P Valadez S, Hartmut Gliemann, Katja Haas-Santo, Christof Wöll, and Roland Dittmeyer. Zif-8 surmof membranes synthesized by au-assisted liquid phase epitaxy for application in gas separation. *Chemie Ingenieur Technik*, 88(11):1798–1805, 2016.

[36] Jin C Tan, Thomas D Bennett, and Anthony K Cheetham. Chemical structure, network topology, and porosity effects on the mechanical properties of zeolitic imidazolate frameworks. *Proceedings of the National Academy of Sciences*, 107(22):9938–9943, 2010.

[37] David Fairen-Jimenez, Stephen A Moggach, Michael T Wharmby, Paul A Wright, Simon Parsons, and Tina Düren. Opening the gate: framework flexibility in zif-8 explored by experiments and simulations. *Journal of the American Chemical Society*, 133(23):8900–8902, 2011.

[38] Surendar R Venna and Moises A Carreon. Highly permeable zeolite imidazolate framework-8 membranes for co2/ch4 separation. *Journal of the American Chemical Society*, 132(1):76–78, 2010.

[39] Uyen PN Tran, Ky KA Le, and Nam TS Phan. Expanding applications of metal- organic frameworks: zeolite imidazolate framework zif-8 as an efficient heterogeneous catalyst for the knoevenagel reaction. *ACS Catalysis*, 1(2):120–127, 2011.

[40] Guang Lu and Joseph T Hupp. Metal- organic frameworks as sensors: a zif-8 based fabry- pérot device as a selective sensor for chemical vapors and gases. *Journal of the American Chemical Society*, 132(23):7832–7833, 2010.

[41] Bo Wang, Adrien P Côté, Hiroyasu Furukawa, Michael O'Keeffe, and Omar M Yaghi. Colossal cages in zeolitic imidazolate frameworks as selective carbon dioxide reservoirs. *Nature*, 453(7192):207–211, 2008.

[42] Claudia Weidenthaler. Pitfalls in the characterization of nanoporous and nanosized materials. *Nanoscale*, 3(3):792–810, 2011.

[43] Daniel Loner. Structure analysis from powder diffraction data. *Powder diffraction: proceedings of the II International School on Powder Diffraction, IACS, Kolkata, India*, pages 1–16, 2002.

[44] Rajamani Krishna and Jasper M van Baten. Hydrogen bonding effects in adsorption of water- alcohol mixtures in zeolites and the consequences for the characteristics of the maxwell- stefan diffusivities. *Langmuir*, 26(13):10854–10867, 2010.

[45] Ma Josephine C Ordonez, Kenneth J Balkus, John P Ferraris, and Inga H Musselman. Molecular sieving realized with zif-8/matrimid® mixed-matrix membranes. *Journal of Membrane Science*, 361(1-2):28–37, 2010.

[46] Dewi W Lewis, A Rabdel Ruiz-Salvador, Ariel Gómez, L Marleny Rodriguez-Albelo, François-Xavier Coudert, Ben Slater, Anthony K Cheetham, and Caroline Mellot-Draznieks. Zeolitic imidazole frameworks: structural and energetics trends compared with their zeolite analogues. *CrystEngComm*, 11(11):2272–2276, 2009.

[47] Norman Colthup. *Introduction to infrared and Raman spectroscopy.* Academic Press, Inc., 1975.

[48] Vladimir G Bordo and Horst-Günter Rubahn. *Optics and spectroscopy at surfaces and interfaces.* John Wiley & Sons, 2005.

[49] Michele R Derrick, Dusan Stulik, and James M Landry. *Infrared spectroscopy in conservation science.* J. Paul Getty Trust, 1999.

[50] Robert G Greenler. Infrared study of adsorbed molecules on metal surfaces by reflection techniques. *The Journal of Chemical Physics*, 44(1):310–315, 1966.

[51] Yue Hu, Hossein Kazemian, Sohrab Rohani, Yining Huang, and Yang Song. In situ high pressure study of zif-8 by ftir spectroscopy. *Chemical Communications*, 47(47):12694–12696, 2011.

[52] Joshua B James and Jerry YS Lin. Thermal stability of zif-8 membranes for gas separations. *Journal of Membrane Science*, 532:9–19, 2017.

[53] Lloyd M Robeson. The upper bound revisited. *Journal of Membrane Science*, 320(1-2):390–400, 2008.

[54] Patricia Gorgojo Alonso. *Desarrollo de materiales laminares porosos para la preparación de membranas híbridas.* PhD thesis, Universidad de Zaragoza, 2010.

[55] Joaquín Coronas and Jesús Santamaria. State-of-the-art in zeolite membrane reactors. *Topics in Catalysis*, 29(1-2):29–44, 2004.

[56] Rolando MA Roque-Malherbe. *Adsorption and diffusion in nanoporous materials.* CRC press, 2007.

[57] Sisir K Roy. *Thermal Physics and Statistical Mechanics.* New Age International, 2001.

[58] Abdullah Alkhudhiri, Naif Darwish, and Nidal Hilal. Membrane distillation: a comprehensive review. *Desalination*, 287:2–18, 2012.

[59] Wenjin Ding. *Simulation-Assisted Design of Polycrystalline Zeolite Catalysts*. PhD thesis, Karlsruher Institut für Technologie, 2016.

[60] Tadashi Uragami. *Science and Technology of Separation Membranes*. John Wiley & Sons, 2017.

[61] JGA Bitter. *Transport mechanisms in membrane separation processes*. Springer Science & Business Media, 1991.

[62] Johannes Holtbrügge, Anna-Katharina Kunze, Alexander Niesbach, Patrick Schmidt, Robin Schulz, Daniel Sudhoff, Mirko Skiborowski, Philip Lutze, and Andrzej Górak. *Reactive and Membrane-Assisted Separations*. Walter de Gruyter GmbH, 2016.

[63] David Farrusseng. *Metal-organic frameworks: applications from catalysis to gas storage*. John Wiley & Sons, 2011.

[64] Warren L McCabe, Julian C Smith, and Peter Harriott. *Unit operations of chemical engineering*. McGraw-Hill chemical engineering series. McGraw Hill, 2005.

[65] JD Seader and Ernest J Henley. *Separation process principles*. John Wiley & Sons, 2006.

[66] Edward N Fuller, Paul D Schettler, and J Calvin Giddings. A new method for prediction of binary gas-phase diffusion coefficients. *Industrial & Engineering Chemistry*, 58(5):18–27, 1966.

[67] Edward L Cussler. *Diffusion: mass transfer in fluid systems*. Cambridge University Press, 2009.

[68] HP Hsieh. *Inorganic membranes for separation and reaction*. Membrane Science and Technology Series, 3. Elsevier Science B.V., 1996.

[69] Weidong He, Weiqiang Lv, and James H Dickerson. Gas diffusion mechanisms and models. In *Gas Transport in Solid Oxide Fuel Cells*, pages 9–17. Springer, 2014.

[70] Suresh K Bhatia, Mauricio R Bonilla, and David Nicholson. Molecular transport in nanopores: a theoretical perspective. *Physical Chemistry Chemical Physics*, 13(34):15350–15383, 2011.

[71] James R Welty, Charles E Wicks, Gregory Rorrer, and Robert E Wilson. *Fundamentals of momentum, heat, and mass transfer*. John Wiley & Sons, 2008.

[72] Rajamani Krishna and Johannes A Wesselingh. The maxwell-stefan approach to mass transfer. *Chemical Engineering Science*, 52(6):861–911, 1997.

[73] Aggelos Doukelis, Kyriakos Panopoulos, Antonios Koumanakos, and Emmanouil Kakaras. *Palladium Membrane Technology for Hydrogen Production, Carbon Capture and Other Applications: Principles, Energy Production and Other Applications*. Woodhead Publishing Series in Energy: Number 68. Elsevier, 2015.

[74] Bernd Kögl and Franz Moser. *Grundlagen der Verfahrenstechnik*. Springer-Verlag Wien GmbH, 1981.

[75] Vitaly Gitis and Gadi Rothenberg. *Ceramic membranes: new opportunities and practical applications*. John Wiley & Sons, 2016.

[76] Rajamani Krishna. Multicomponent surface diffusion of adsorbed species: a description based on the generalized maxwell-stefan equations. *Chemical Engineering Science*, 45(7):1779–1791, 1990.

[77] Roland Dittmeyer and Gerhard Emig. chapter Simultaneous heat and mass transfer and chemical reaction, pages 1727–1784. Wiley-VCH Verlag GmbH, 2008.

[78] Jolinde M van de Graaf, Freek Kapteijn, and Jacob A Moulijn. Modeling permeation of binary mixtures through zeolite membranes. *AIChE journal*, 45(3):497–511, 1999.

[79] József Tóth. *Adsorption: theory, modeling, and analysis*. Marcel Dekker, Inc., 2002.

[80] Alan L Myers and John M Prausnitz. Thermodynamics of mixed-gas adsorption. *AIChE Journal*, 11(1):121–127, 1965.

[81] Jiahui Chen, Leslie S Loo, and Kean Wang. An ideal absorbed solution theory (iast) study of adsorption equilibria of binary mixtures of methane and ethane on a templated carbon. *Journal of Chemical & Engineering Data*, 56(4):1209–1212, 2011.

[82] Krista S Walton and David S Sholl. Predicting multicomponent adsorption: 50 years of the ideal adsorbed solution theory. *AIChE Journal*, 61(9):2757–2762, 2015.

[83] Alessandro Erto, Amedeo Lancia, and Dino Musmarra. A real adsorbed solution theory model for competitive multicomponent liquid adsorption onto granular activated carbon. *Microporous and Mesoporous Materials*, 154:45–50, 2012.

[84] Michael Sakuth, Johannes Meyer, and Jürgen G Gmehling. Measurement and prediction of binary adsorption equilibria of vapors on dealuminated y-zeolites (day). *Chemical Engineering and Processing: Process Intensification*, 37(4):267–277, 1998.

[85] Jörg Kärger and Sergey Vasenkov. *Handbook of Zeolite Science and Technology*, chapter Diffusion in zeolites, pages 446–548. Marcel Dekker, Inc., 2003.

[86] Ulrike Böhme, Benjamin Barth, Carolin Paula, Andreas Kuhnt, Wilhelm Schwieger, Alexander Mundstock, Jürgen Caro, and Martin Hartmann.

Ethene/ethane and propene/propane separation via the olefin and paraffin selective metal–organic framework adsorbents cpo-27 and zif-8. *Langmuir*, 29(27):8592–8600, 2013.

[87] Brian R Pimentel and Ryan P Lively. Enabling kinetic light hydrocarbon separation via crystal size engineering of zif-8. *Industrial & Engineering Chemistry Research*, 55(48):12467–12476, 2016.

[88] Wei Su, Ai Zhang, Yan Sun, Meng Ran, and Xiaojing Wang. Adsorption properties of c2h4 and c3h6 on 11 adsorbents. *Journal of Chemical & Engineering Data*, 62(1):417–421, 2017.

[89] Joshua B James, Jun Wang, Lie Meng, and Jerry YS Lin. Zif-8 membrane ethylene/ethane transport characteristics in single and binary gas mixtures. *Industrial & Engineering Chemistry Research*, 56(26):7567–7575, 2017.

[90] Yichang Pan, Tao Li, Gabriella Lestari, and Zhiping Lai. Effective separation of propylene/propane binary mixtures by zif-8 membranes. *Journal of membrane science*, 390-391:93–98, 2012.

[91] Helge Bux, Christian Chmelik, Rajamani Krishna, and Jürgen Caro. Ethene/ethane separation by the mof membrane zif-8: molecular correlation of permeation, adsorption, diffusion. *Journal of membrane science*, 369(1-2):284–289, 2011.

[92] Sonia Aguado, Charles-Henri Nicolas, Virginie Moizan-Baslé, Carlos Nieto, Hedi Amrouche, Nicolas Bats, Nathalie Audebrand, and David Farrusseng. Facile synthesis of an ultramicroporous mof tubular membrane with selectivity towards co2. *New Journal of Chemistry*, 35(1):41–44, 2011.

[93] Helge Bux, Fangyi Liang, Yanshuo Li, Janosch Cravillon, Michael Wiebcke, and Jürgen Caro. Zeolitic imidazolate framework membrane with molecular sieving properties by microwave-assisted solvothermal synthesis. *Journal of the American Chemical Society*, 131(44):16000–16001, 2009.

[94] Miral Shah, Hyuk T Kwon, Vu Tran, Sonny Sachdeva, and Hae-Kwon Jeong. One step in situ synthesis of supported zeolitic imidazolate framework zif-8 membranes: Role of sodium formate. *Microporous and Mesoporous Materials*, 165:63–69, 2013.

[95] Yichang Pan, Yunyang Liu, Gaofeng Zeng, Lan Zhao, and Zhiping Lai. Rapid synthesis of zeolitic imidazolate framework-8 (zif-8) nanocrystals in an aqueous system. *Chemical Communications*, 47(7):2071–2073, 2011.

[96] Kang Huang, Ziye Dong, Qianqian Li, and Wanqin Jin. Growth of a zif-8 membrane on the inner-surface of a ceramic hollow fiber via cycling precursors. *Chemical Communications*, 49(87):10326–10328, 2013.

[97] Koji Kida, Kosuke Fujita, Tomoko Shimada, Shunsuke Tanaka, and Yoshikazu Miyake. Layer-by-layer aqueous rapid synthesis of zif-8 films on a reactive surface. *Dalton Transactions*, 42(31):11128–11135, 2013.

[98] Richard D Noble and S Alexander Stern. *Membrane separations technology: principles and applications*. Membrane Science and Technology Series, 2. Elsevier Science B.V., 1995.

[99] Michael C McCarthy, Victor Varela-Guerrero, Gregory V Barnett, and Hae-Kwon Jeong. Synthesis of zeolitic imidazolate framework films and membranes with controlled microstructures. *Langmuir*, 26(18):14636–14641, 2010.

[100] Yichang Pan and Zhiping Lai. Sharp separation of c2/c3 hydrocarbon mixtures by zeolitic imidazolate framework-8 (zif-8) membranes synthesized in aqueous solutions. *Chemical Communications*, 47(37):10275–10277, 2011.

[101] Yichang Pan, Bo Wang, and Zhiping Lai. Synthesis of ceramic hollow fiber supported zeolitic imidazolate framework-8 (zif-8) membranes with high hydrogen permeability. *Journal of membrane science*, 421-422:292–298, 2012.

[102] Hang Yin, Hyungmin Kim, Jungkyu Choi, and Alex CK Yip. Thermal stability of zif-8 under oxidative and inert environments: A practical perspective on using zif-8 as a catalyst support. *Chemical Engineering Journal*, 278:293–300, 2015.

[103] GK Williamson and WH Hall. X-ray line broadening from filed aluminium and wolfram. *Acta metallurgica*, 1(1):22–31, 1953.

[104] Chen Zhang, Ryan P Lively, Ke Zhang, Justin R Johnson, Oguz Karvan, and William J Koros. Unexpected molecular sieving properties of zeolitic imidazolate framework-8. *The journal of physical chemistry letters*, 3(16):2130–2134, 2012.

[105] Stephen A Moggach, Thomas D Bennett, and Anthony K Cheetham. The effect of pressure on zif-8: Increasing pore size with pressure and the formation of a high-pressure phase at 1.47 gpa. *Angewandte Chemie International Edition*, 48(38):7087–7089, 2009.

[106] Yue Hu, Zhenxian Liu, Jun Xu, Yining Huang, and Yang Song. Evidence of pressure enhanced co2 storage in zif-8 probed by ftir spectroscopy. *Journal of the American Chemical Society*, 135(25):9287–9290, 2013.

[107] Víctor MA Melgar, Jinsoo Kim, and Mohd R Othman. Zeolitic imidazolate framework membranes for gas separation: A review of synthesis methods and gas separation performance. *Journal of Industrial and Engineering Chemistry*, 28:1–15, 2015.

[108] Bin Li, Hui-Min Wen, Wei Zhou, and Banglin Chen. Porous metal–organic frameworks for gas storage and separation: what, how, and why? *The journal of physical chemistry letters*, 5(20):3468–3479, 2014.

[109] Rahul Banerjee, Anh Phan, Bo Wang, Carolyn Knobler, Hiroyasu Furukawa, Michael O'keeffe, and Omar M Yaghi. High-throughput synthesis of zeolitic imidazolate frameworks and application to co2 capture. *Science*, 319(5865):939–943, 2008.

[110] E Wicke and R Kallenbach. Surface diffusion of carbon dioxide in activated charcoals. *Kolloid Z*, 97:135, 1941.

[111] Jörg Kärger, Douglas M Ruthven, and Doros N Theodorou. *Diffusion in nanoporous materials*. John Wiley & Sons, 2012.

[112] Scott M Auerbach, Kathleen A Carrado, and Prabir K Dutta. *Handbook of zeolite science and technology*. Marcel Dekker, Inc., 2003.

[113] Eric G Derouane, Valentin Parmon, Francisco Lemos, and Fernando R Ribeiro. *Principles and Methods for Accelerated Catalyst Design and Testing*. NATO Science Series. Springer Science & Business Media, 2002.

[114] Antonie J Burggraaf, Zeger AEP Vroon, Klaas Keizer, and Hendrik Verweij. Permeation of single gases in thin zeolite mfi membranes. *Journal of membrane science*, 144(1-2):77–86, 1998.

[115] Jolinde M van de Graaf, Freek Kapteijn, and Jacob A Moulijn. Methodological and operational aspects of permeation measurements on silicalite-1 membranes. *Journal of Membrane Science*, 144(1-2):87–104, 1998.

[116] Wei F Guo, Tai-Shung Chung, and Takeshi Matsuura. Pervaporation study on the dehydration of aqueous butanol solutions: a comparison of flux vs. permeance, separation factor vs. selectivity. *Journal of Membrane Science*, 245(1-2):199–210, 2004.

[117] Meha Rungta, Chen Zhang, William J Koros, and Liren Xu. Membrane-based ethylene/ethane separation: The upper bound and beyond. *AIChE Journal*, 59(9):3475–3489, 2013.

[118] Jolinde M van de Graaf, Freek Kapteijn, and Jacob A Moulijn. Permeation of weakly adsorbing components through a silicalite-1 membrane. *Chemical engineering science*, 54(8):1081–1092, 1999.

[119] DD Do and HD Do. Cooperative and competitive adsorption of ethylene, ethane, nitrogen and argon on graphitized carbon black and in slit pores. *Adsorption*, 11(1):35–50, 2005.

[120] Maaike C Kroon and Lourdes F Vega. Selective paraffin removal from ethane/ethylene mixtures by adsorption into aluminum methylphosphonate-α: a molecular simulation study. *Langmuir*, 25(4):2148–2152, 2009.

[121] Ying Wu, Huiyong Chen, Defei Liu, Yu Qian, and Hongxia Xi. Adsorption and separation of ethane/ethylene on zifs with various topologies: Combining gcmc simulation with the ideal adsorbed solution theory (iast). *Chemical Engineering Science*, 124:144–153, 2015.

[122] Xijun Hu and Duong D Do. Comparing various multicomponent adsorption equilibrium models. *AIChE journal*, 41(6):1585–1592, 1995.

[123] Cory M Simon, Berend Smit, and Maciej Haranczyk. pyiast: Ideal adsorbed solution theory (iast) python package. *Computer Physics Communications*, 200:364–380, 2016.

[124] Angelo Basile. *Handbook of membrane reactors: fundamental materials science, design and optimisation.* Woodhead Publishing Series in Energy: Number 55. Woodhead Publishing, 2013.

[125] William J Koros, Yi H Ma, and T Shimidzu. Terminology for membranes and membrane processes (iupac recommendations 1996). *Pure and Applied Chemistry*, 68(7):1479–1489, 1996.

[126] Martin Knudsen. Die gesetze der molekularströmung und der inneren reibungsströmung der gase durch röhren. *Annalen der Physik*, 333(1):75–130, 1909.

[127] Martin Knudsen and Willard J Fisher. The molecular and the frictional flow of gases in tubes. *Physical Review (Series I)*, 31(5):586–588, 1910.

[128] Marian von Smoluchowski. Zur kinetischen theorie der transpiration und diffusion verdunnter gase. *Annalen der Physik*, 338(16):1559–1570, 1910.

[129] Marian von Smoluchowski. Zur theorie der wärmeleitung in verdünnten gasen und der dabei auftretenden druckkräfte. *Annalen der Physik*, 340(10):983–1004, 1911.

[130] Wenjin Ding, Michael Klumpp, Seungcheol Lee, Stephanie Reuß, Shaeel A Al-Thabaiti, Peter Pfeifer, Wilhelm Schwieger, and Roland Dittmeyer. Simulation of one-stage dimethyl ether synthesis over a core-shell catalyst. *Chemie Ingenieur Technik*, 87(6):702–712, 2015.

[131] Rajamani Krishna. A unified approach to the modelling of intraparticle diffusion in adsorption processes. *Gas separation & purification*, 7(2):91–104, 1993.

[132] Thomas Melin and Robert Rautenbach. *Membranverfahren: Grundlagen der Modul-und Anlagenauslegung*. Springer-Verlag, 2007.

[133] Arvind Kapoor, James A Ritter, and Ralph T Yang. An extended langmuir model for adsorption of gas mixtures on heterogeneous surfaces. *Langmuir*, 6(3):660–664, 1990.

[134] Ralph T Yang. *Gas separation by adsorption processes*. Butterworths Series in Chemical Engineering. Butterworth Publishers, 1987.

[135] M Douglas LeVan and Theodore Vermeulen. Binary langmuir and freundlich isotherms for ideal adsorbed solutions. *The Journal of Physical Chemistry*, 85(22):3247–3250, 1981.

[136] Jianwen Jiang. *Metal-Organic Frameworks: Materials Modeling towards Engineering Applications*. CRC Press, 2015.

[137] Rutger A van Santen and Matthew Neurock. *Molecular heterogeneous catalysis: a conceptual and computational approach*. John Wiley & Sons, 2006.

Publications

Parts of this dissertation have already been published in:

Elvia P Valadez S, Hartmut Gliemann, Katja Haas-Santo, Christof Wöll, and Roland Dittmeyer. ZIF-8 SURMOF membranes synthesized by Au-assisted liquid phase epitaxy for application in gas separation. *Chemie Ingenieur Technik*, 88(11):1798–1805, 2016.